Geographies of
Postcolonialism

Geographies of Postcolonialism

Spaces of Power and Representation

Joanne P. Sharp

Los Angeles • London • New Delhi • Singapore • Washington DC

SAGE Publications Ltd
1 Oliver's Yard
55 City Road
London EC1Y 1SP

SAGE Publications Inc.
2455 Teller Road
Thousand Oaks, California 91320

SAGE Publications India Pvt Ltd
B 1/I 1 Mohan Cooperative Industrial Area
Mathura Road, New Delhi 110 044

SAGE Publications Asia-Pacific Pte Ltd
33 Pekin Street #02-01
Far East Square
Singapore 048763

Library of Congress Control Number: 2008906723

British Library Cataloguing in Publication data

A catalogue record for this book is available from the British Library

ISBN 978-1-4129-0778-1
ISBN 978-1-4129-0779-8 (pbk)

Typeset by C&M Digitals (P) Ltd, Chennai, India
Printed in Great Britain by The Cromwell Press Ltd, Trowbridge, Wiltshire
Printed on paper from sustainable resources

For Margaret Johnstone

CONTENTS

LIST OF FIGURES AND TABLES

Figures

Tables

ACKNOWLEDGEMENTS

This book is based around an Honours course that I have taught at the University of Glasgow over the last ten years. When I was first putting the course together, Jim Duncan had just arrived at Cambridge and was also writing a new course for his students. In those days of frantically pulling notes and reading lists together, various bits and pieces of lectures were emailed back and forth, and so parts of this book bear the imprint of Jim's scholarship, for which it has benefited greatly. Going back further, it was Linda Alcoff's exceptional graduate course on Postcolonialism that inspired me to pursue this topic further, and to Beverley Allen I have a debt of gratitude for introducing me to *The Satanic Verses* and *Xala*. More recently, John Briggs has encouraged me to confront my postcolonial assumptions in places outside of Western academy, for which I am eternally grateful.

Over the years teaching the course I have been lucky enough to have taught a number of excellent students who have engaged with, and challenged, this material (and interrupted the flow of my lectures) with enthusiasm and humour. I would like to thank them all, but especially Andy McMillan, Allan Lafferty, Martin Muir and Olly Zanetti. Thanks also go to Geraldine Perriam who generously read through an earlier draft of the book, and who has inspired me with her discussions of writing; and to John Moore of Glasgow University Library who helped me to source some of the images. Finally, thanks to Robert Rojek and Sarah-Jayne Boyd at Sage for their encouragement, support and patience during the production of this book.

This book is dedicated to my grandmother, Margaret Johnstone, who died in February 2007 when I was completing the manuscript. While there is a lot here she would probably disagree with, I miss the opportunity to hear her say so.

PUBLISHER'S ACKNOWLEDGEMENTS

The author and publishers wish to thank the following for the permission to use copyright material:

1.1 Medieval map depicting 'monstrous races' around the margins of known geographical space. Source: Mappae Mundi: die ältesten Weltkarten, (Jos Roth'Sche Verlagshandlung: Stuttgart) 1895

1.4 Eugène Delacroix, *Fanatics of Tangier* (Les convulsionnaires de Tanger). Oil on Canvas 1837–1838. Courtesy of Minneapolis Institute of Arts, Bequest of J. Jerome Hill

1.5 *Dance of the Almeh*, Gerome, 1863. Oil on wood panel, 19 3/4 x 32 inches (50.2 x 8.3cm). Courtesy of the Dayton Art Institute. Gift of Mr. Robert Badenhop, 1951.15

1.6 *Gateway to the Great Temple at Balbec*, Roberts, 1841. Courtesy of The Royal Academy of Arts, London

2.1 Education Euder, (1856), *Alexander von Humboldt and Aimé Bonpland in Urwald*. Berlin-Brandenburgische Akademie der Wissenschaften.

2.2 Jan van der Straet's *America*. Courtesy of The National Gallery of Art, Washington

2.4 Cairo street scene at the Exposition Universelle, Paris, 1889. Source: Glasgow University Library

2.5 *Carry on up the Khyber* montage. Source: Peter Rogers Production

3.1 Bentham's panopticon. Source: Foucault, Michael, *Discipline and Punishment*, 1975

3.2 Schematic diagram of a coffee plantation, mid-nineteenth century, from Duncan, J.S. (2002) 'Embodying Colonialism?: Domination and Resistance in nineteenth century Ceylonese Coffee Plantations.' This article was published in the *Journal of Historical Geography*, 28 (3): 317–38. © Elsevier

3.3 Contrasting Delhi streetscapes. Photo credits: John Briggs

5.1 Mecca-Cola. Source: www.mecca-cola.com

5.2 Indiana Jones as saviour © Lucasfilm Ltd

5.4 Benetton advert © Copyright 1990 Benetton Group S.p.A. – Photo: Oliviero Toscani

5.5 Drop the Debt advert. Courtesy of the Jubilee Debt Campaign

5.6 Anti-Coke graffiti. Source: Andrew McMillan

5.7 Internet access rates, 2000. Courtesy of Jeremy Crampton. Source: *The Political Mapping of Cyberspace* (Edinburgh University Press, 2003).

6.1 Mehdy Kavousi © Vincent Jannink/epa/Corbis

6.2 'The law vs. Ayook' from 'Colonialism On Trial' by D.H. Monet and Skanu'u (1992). Courtesy of D. H. Monet, Canada | info@donmonet.ca

6.3 Orientalist image of the veil. *Indigène Du Caire*, G. Lekegian and Cie. Source: Photography Collection, Miriam and Ira D. Wallach Division of Art, Prints and Photographs, The New York Public Library Astor, Lenox and Tilden Foundations.

6.4 Veils as fashion. Courtesy of © www.TheHijabShop.com

6.5 Covers of Fanon's *Black Skins, White Masks*. Courtesy of Pluto Books, www.plutobooks.com

6.6 Images of women from *The Battle of Algiers*. Source: Criterion Studio

8.1 Bedouin women shaking the branches of a tree to dislodge leaves for their goats. Courtesy of Irina Springuel

While every effort has been made to trace the owners of copyright material, in a few cases this has proved impossible and we take this opportunity to offer our apologies to any copyright holder whose rights we have unwittingly infinged.

INTRODUCTION

Along with language, it is geography – especially for the displaced form of departures, arrivals, farewells, exile, nostalgia, homesickness, belonging, and travel itself – that is at the core of my memories of those early years.

[…]

… the overriding sensation I had was of always being out of place. Thus it took me about fifty years to become accustomed to, or, more exactly, to feel less uncomfortable with, 'Edward,' a foolishly English name yoked forcibly to the unmistakably Arabic family name Said. True my mother told me I had been named Edward after the Prince of Wales, who cut so fine a figure in 1935, the year of my birth, and Said was the name of various uncles and cousins. But the rationale of my name broke down both when I discovered no grandparents named Said and when I tried to connect my fancy English name with its Arabic partner. For years, depending on the exact circumstances, I would rush past 'Edward' and emphasize 'Said'; at other times I would do the reverse, or connect these two to each other so quickly that neither would be clear. The one thing I could not tolerate, but very often would have to endure, was the disbelieving, and hence undermining, reaction: Edward? Said?

The travails of bearing such a name were compounded by an equally unsettling quandary when it came to language. I have never known which language I spoke first, Arabic or English, or which one was really mine beyond any doubt. What I do know, however, is that the two have always been together in my life, one resonating in the other, sometimes ironically, sometimes nostalgically, most often each correcting, and commentating on, the other.

Edward Said (1999) Out of Place: A Memoir, pp. xvi, 3–4

I am a border woman. I grew up between two cultures, the Mexican (with a heavy Indian influence) and the Anglo (as a member of a colonized people in our own territory). I have been straddling that *tejas*-Mexican border, and others, all my life. It's not a comfortable territory to live in, this place of contradictions. Hatred, anger and exploitation are the prominent features of this landscape.

However, there have been compensations for this *mestiza*, and certain joys. Living on borders and in margins, keeping intact one's shifting and multiple identity and integrity, is like trying to swim in a new element, an 'alien' element. There is an exhilaration in being a participant in the further evolution of humankind, in being 'worked' on. I have the sense that certain 'faculties' – not just in me but in every border resident, colored or non-colored – and dormant areas of consciousness are being activated, awakened. Strange, huh? Any yes, the 'alien' element has become familiar – never comfortable, not with society's clamor to uphold the old, to rejoin the flock, to go with the herd. No, not comfortable but home.

Gloria Anzaldúa (1987) Borderlands/La Frontera: The New Mestiza, p.vii

I was born in Palo Alto, California, into the lap of an Iranian diaspora community awash in nostalgia and longing for an Iran many thousands of miles away. A girl, raised on the distorting myths of exile. I imagined myself a Persian princess, estranged from my homeland – a place of light, poetry, and nightingales – by a dark, evil force called the Revolution. I borrowed the plot from *Star Wars*, convinced it told Iran's story. Ayatolla Khomeini was Darth Vader. Tromping around suburban California, I lived out this fantasy. There must be some supernatural explanation, I reasoned, for the space landing of thousands of Tehranis to a world of vegan smoothies and Volvos, chakras, and Tupak.

Growing up, I had no doubt that I was Persian. Persian like a fluffy cat, a silky carpet – a vaguely Oriental notion belonging to history, untraceable on a map. It was the term we insisted on using at the time, embarrassed by any association with Iran, the modern country, the hostage-taking Death Star. Living a myth, a fantasy, made it easier to be Iranian in America.

As life took its course, as I grew up and went to college, discovered myself, and charted a career, my Iranian sense of self remained intact. But when I moved to Tehran in 2000 – pleased with my pluckiness, and eager to prove myself as a young journalist – it, along with the fantasies, dissolved. Iran, as it turned out, was not the Death Star, but a country where people voted, picked their noses, and ate French fries. Being a Persian girl in California, it turned out, was like, a totally different thing than being a young Iranian woman in the Islamic Republic of Iran.

Azadeh Moaveni (2005) Lipstick Jihad:
A Memoir of Growing up Iranian in America and
American in Iran, p. vii

These three autobiographies are all the products of a globalised world where home and identity are complex constructs emerging from cultural contact, mixing and mobility. Each experience is post-colonial, of a world after the colonial period in which these three people – like each of us – are created by the powers, connections and imaginations that were written into the world during Europeans' first explorations of the world and the making and remaking of these geographies ever since. Said's identity was directed by Middle Eastern politics and memories of place and belonging, and as a Palestinian his sense of self and identity was entangled with exile, dispossession and displacement; Anzaldúa's experiences are fundamentally structured by the size and power of the US-side of the borderlands she inhabits; and Moaveni's sense of self has been formed through her inherited imagined geographies of her origins in the exotic land of 'Persia'. Each is a postcolonial subject constituted through real and imagined geographical processes and identities, through ongoing conflicts, stereotyping and the fantasising of different parts of the world.

The geographies that make up these people's experiences and identities reflect the fluidity of our contemporary globalised world, at the same time as recognising the continued existence of differences and barriers (of outside, exotic, alien …) that were formed in previous periods and continue to shape our geographical imaginations. These identities are the result of cultural mixing and hybridisation – the processes of globalisation we hear so much about – but these are not free combinations. Certain parts of the mix have greater power to influence the direction of change (the power of English over Arabic as a global language; the greater constraints of movement over the US-Mexican border for Mexicans over citizens of the USA; the power of the western imagination to conjure up the exotic east). Postcolonial geographies then are this ambiguous mix of the fluid and the unchanging that shapes the identities of people like Said, Anzaldúa and Moaveni … and all of us. Postcolonialism is structured through geographies of imagination, knowledge and power, and it is these geographies that will be at the heart of this book.

WHAT IS POST(-)COLONIALISM? THE IMPORTANCE OF A HYPHEN

Since the early 1980s, postcolonialism has developed a body of writing that attempts to shift the dominant ways in which relations between western and non-western people and their worlds are viewed … postcolonialism seeks to intervene, to force its alternative knowledges into the power structures of the west as well as the non-west. It seeks to change the way people think, the way they behave, to produce a more just and equitable relation between the different peoples of the world. (Young, 2003: 2)

To understand how it is that relations have formed between western and non-western peoples, it is necessary for postcolonialism to have a historical vision. Before we can move on to post(-)colonialism, we need a definition of colonialism.

Colonialism always assumes the physical occupation of one land by peoples associated with another place. The colonists do not simply remove resources and wealth from the new land (as is the case with forms of **imperialism**) but actually occupy the territory, building settlements, and often also agriculture and industry. There have been many instances of colonialism through human history, for instance the Roman Empire witnessed colonies from Britain through to the Mediterranean and into the Middle East. However, in this book we will only be looking at the period of European colonialism which was initiated with the 'Age of Exploration' where Europeans started to explore new continents, and which reached its high point in the nineteenth century.

This form of colonialism was distinct not only because of its unprecedented scale but also because of its establishment alongside a specific form of rational knowledge (called the **European Enlightenment**) which saw science emerge as the most important form of knowledge, and also witnessed the rise of mercantile capitalism which was driven both by the possibilities available in the new lands and also by the rise of scientific knowledge which objectified the world into measurable land to be owned and resources to be exploited for the colonisers' use. Thus, the way that European colonists came to know the world has been highly influential. The combination of scientific knowledge and capitalism within the context of superiority provided the framework through which the new lands and peoples became known to the Europeans and subsequently became the basis for European control of them. In many cases, this knowledge also became the way in which the peoples the Europeans ruled came to know themselves.

There are two different ways in which post(-)colonialism is understood as a term, differentiated by the use of a hyphen, although different authors have varying interpretations of what the hyphen does mean. Blunt and McEwan (2002: 3) argue that the '"post" of "postcolonialism" has two meanings, referring to a temporal aftermath – a period of time *after* colonialism – and a critical aftermath – cultures, discourses and critiques that lie *beyond*, but remain closely influenced by, colonialism'. Thus, although the definitions are clearly related, the differences in meaning can be drawn out as follows.

Post-colonialism

When the hyphen is used in the term, it refers to the common-sensical definition of post-colonialism as the period following independence from colonising powers. Thus, it is both a geographical term (particular countries are post-colonial) and a historical period. Some see this definition as problematic, as it over-emphasises the break. In his analysis of the (geo)politics of contemporary Afghanistan, Palestine and Iraq, Derek Gregory (2004) adopts the term 'colonial present' to emphasise further still the continuities in imagined geographies between the past and present.

Postcolonialism

However, postcolonialism is also a critical approach to analysing colonialism and one that seeks to offer alternative accounts of the world. The term is written without the hyphen to recognise the problems with the first, and more conventional, use of the concept. This recognises clear tensions within this term. For, while it is a concept that seeks to challenge colonialism and the values and meanings it depended upon, the name ties it strongly to what went before. Rather than being a positive concept it is a negative one: it is *not* colonialism. As Anne McClintock (1995: 11) has put it, postcolonialism 'confers on colonialism the prestige of history proper ... the world's multitudinous cultures are marked, not positively by what distinguishes them, but by subordinate, retrospective relation to linear, European time'. This second definition seeks to play on the ambiguity of the concept, recognising continuities from the colonial period as well as breaks from it – and also recognising that while states might be physically decolonised, this does not mean that other effects of the colonial period have all disappeared. This is because postcolonialism also represents a shift from a form of analysis based solely around politics and economics (again the conventional way of understanding the impacts of colonialism) to consider instead the importance of the cultural products of colonialism, particularly the ways of knowing the world that emerged.

Thus, postcolonialists have argued that while political, and to a less extent economic, decolonisation might have occurred with independence, cultural decolonisation – what some call the decolonisation of the mind – has been a much more difficult process. Western values, science, history, geography and culture were privileged during colonialism as ways in which the colonisers came to know the places and peoples they colonised. However, as these knowledges and values were insinuated through institutions of education, governance and media, they also became (to a greater or lesser extent) the ways in which the colonised came to know themselves. The internalisation of a set of values and ways of knowing the world is much more difficult to overturn than the physical rule of colonial regimes, postcolonial theorists would argue.

Thus, postcolonialism is an analysis and critique of the ways in which western knowledge systems have come to dominate. It is a form of analysis that is focused around cultural productions in order that, as well as looking at the ways in which the world came to be represented in the formal documents of explorers, educators and as governors, it also looks at novels, songs, art, movies and advertising as forms of knowledge about the world, and as ways in which this knowledge is communicated. As we shall see later, however, postcolonialism is also a more positive project which seeks to recover alternative ways of knowing and understanding – often talked of in terms of 'other voices' – in order to present alternatives to dominant western constructs.

POSTCOLONIAL GEOGRAPHIES

Postcolonial theories seem to be very geographical in that the language used talks about spaces, centres, peripheries and borders. There were distinctive geographies of colonialism, in terms of the different ways in which colonial policy was practised across the world. The way that the British treated Indians in South Asia was different from French policies in the Middle East. Both differed from the ways in which colonial administrators ran countries in sub-Saharan Africa. Such differences were the result of national approaches to colonialism, perceptions of the environments and the natives who inhabited these places, and even, to an extent, the influence of the individuals who had responsibility for rule. Despite this, postcolonial theory has suggested a coherence in conceptual approach that transcended these differences. What this means is that in this book we will be looking at the geographical outcomes of colonial and postcolonial processes (the influence on the landscape, representations of place and so on) rather than comparing the practices of colonialism and the postcolonial response to these in different countries. While this loses the historical detail of how specific colonial representations and policies played out in different places, it does allow us to look at the continuities in the construction of colonial knowledge (and resistance to it) which transcend conventional regional geographies (and the ways in which these processes continue into post-colonial practice).

Geography is very important to postcolonialism. On the whole, postcolonial theory has been dominated by scholars from the discipline of literature. Their focus has been on the texts of colonialism in terms of the books written by travellers, academics, colonial administrators, anti-colonial resisters, politicians and novelists, amongst others. These are important texts, all the more so because previous approaches to colonialism ignored such sources. The words on the pages of these texts have had a great influence on how we see the world and the various connections between its different parts. However, these texts are perhaps ideals – how colonial societies *should* be organised in an ideal world, the maps of colonial spaces or treatises for how the post-colonial world order *should* play out. But when texts turn into practices, all sorts of other things come into play. Most importantly, there are all sorts of questions of translation: how will texts translate into other languages and be read by those with different cultural backgrounds? How will buildings or agricultural practices translate into environments that are very different to those dominant in the countries where the texts were written? How well will colonial administrators or development workers be able to translate their instructions into day-to-day practice (will they be distorted by ambition, corruption or misunderstandings)? And how well did the natives understand the intentions of colonial practice? They may have believed in them, they may have gone along with them, they might have actively resisted them – or perhaps they just failed to understand what was intended. Each had consequences for the ways in which the colonial texts were translated into real outcomes. A geographical version of postcolonialism is attentive to the ways in which texts are changed as they are translated into practice in particular places around the world.

STRUCTURE OF THE BOOK

I have split the book into three sections, following the first three terms introduced at the beginning of this chapter:

1 *Colonialisms* will consider the ways in which understandings of the rest of the world were incorporated into European knowledge, from the period prior to exploration of the lands beyond Europe's boundaries until the present. We will see how formal knowledge of the world was collected, how this was disseminated through society via education and popular culture, and how the knowledge of this world was translated (or mistranslated) into practice.
2 *Post-colonialisms* stresses the continuities existing between the colonial to the post-colonial periods. This section will consider the cultural similarities and differences that have emerged since the end of the colonial period, looking at the rise of the 'Third World', and development and globalisation as important post-colonial processes.
3 *Postcolonialisms* will think about postcolonialism as a critical theoretical project which challenges western assumptions, stereotypes and ways of knowing and offers its own alternatives. It will also look at the extent to which postcolonialism runs through cultural productions in wider society, and will finish up by examining the relevance of postcolonialism to some of the big questions about poverty and inequality faced in the world today.

As already indicated, one of the goals of postcolonialism is to include voices that have been previously excluded from academic discussions. Postcolonial writers tend to challenge the presentation of singular narratives and instead seek to include multiple voices in their works. Thus, in this book, alongside the story that I am telling about postcolonial geographies, are a series of boxes that include direct and sustained quotations from other authors, both academic and popular, so that you can see how others articulate the issues discussed in each section. These include extracts from the work of travel writers, academics, politicians, novelists and others writing at the times and, sometimes, in the places we are discussing. This means you will have the opportunity to read the original sources alongside my interpretation of them – you may not always agree with my version! I hope that you will not stop at reading these excerpts: although I have attempted to choose passages that represent these other texts well, I hope that these little tasters will encourage you to seek out the originals and read further.

When we are thinking about how the world is represented, when we think about the sources from which we each get our understandings of the world around us, we cannot only look to written sources but must also closely examine images – whether these are the paintings of nineteenth century Orientalists, film stills, illustrations from the *National Geographic*, or advertising images. Thus, there is a lot of illustrative material placed alongside my argument, like the text boxes mentioned above. Please give these more than a passing glance – think about how they work; why the artist or photographer sought to create that particular image; how the meaning of the image might change over time, in different places, depending upon where the image was placed.

At the end of each chapter I have suggested a few sources for you to read up to find out more about particular issues. While these tend to be academic sources, I have also included films and works of fiction, both of which are important sources for finding out about the postcolonial world we inhabit.

Further reading

On postcolonial geography

Blunt, A. and McEwan, C. (eds) (2002) *Postcolonial Geographies*. London: Continuum.

Blunt, A. and Rose, G. (eds) (1994) *Writing Women and Space: Colonial and Postcolonial Geographies*. New York: Guilford Press.

Gregory, D. (2004) *The Colonial Present*. Oxford: Blackwell.

Jacobs, J. (1996) *Edge of Empire*. London: Routledge.

Sidaway, J. (2000) 'Postcolonial geographies: an exploratory essay', *Progress in Human Geography, 24* (4): 591–612.

PART I

COLONIALISMS

Before engaging in any critical analysis of them, it is important to understand how geographies of the rest of the world were established in European culture prior to and during the colonial period. For instance, what did Europeans believe lay beyond the boundaries of known space before they set out to explore the unknown? This is significant, because people do not see the world entirely as it is, but always through the distortions of cultural values and expectations. Once new places and peoples were discovered, how were they incorporated into existing frameworks of knowing and understanding, and how were these knowledges challenged and changed by exploration? We will consider these issues in Chapter 2, and will discuss the most influential work on this topic, a book that many have argued marked the establishment of postcolonialism as an intellectual approach, Edward Said's *Orientalism*. In Chapter 3 we will see how European knowledges of the rest of the world came to power by looking not only at the rise of academic knowledge of new places and peoples, but also the ways in which these knowledges were popularised through travel accounts, advertising and World's Fairs. In the final chapter in this section, we will examine how European knowledges were made material in the remaking of colonial landscapes. Here we see things move from the realm of ideas into practices and the very physicality of the landscape, reinforcing the central argument of postcolonialism of the central importance of cultural values and knowledge to the enduring power of colonialism.

1 IMAGINING THE WORLD

And from that other coast of Chaldea, toward the south, is Ethiopia, a great country that stretcheth to the end of Egypt. Ethiopia is departed in two parts principal, and that is in the east part and in the meridional part; the which part meridional is clept Mauritania; and the folk of that country be black enough and more black than in the tother part, and they be clept Moors. In that part is a well, that in the day it is so cold, that no man may drink thereof; and in the night it is so hot, that no man may suffer his hand therein. And beyond that part, toward the south, to pass by the sea Ocean, is a great land and a great country; but men may not dwell there for the fervent burning of the sun, so is it passing hot in that country.

In Ethiopia all the rivers and all the waters be trouble, and they be somedeal salt for the great heat that is there. And the folk of that country be lightly drunken and have but little appetite to meat. And they have commonly the flux of the womb. And they live not long. In Ethiopia be many diverse folk; and Ethiope is clept Cusis. In that country be folk that have but one foot, and they go so blyve that it is marvel. And the foot is so large, that it shadoweth all the body against the sun, when they will lie and rest them. In Ethiopia, when the children be young and little, they be all yellow; and, when that they wax of age, that yellowness turneth to be all black. In Ethiopia is the city of Saba, and the land of the which one of the three kings that presented our Lord in Bethlehem, was king of.

John Mandeville, Chapter XVII: 'Of the land of Job; and of his age. Of the array of men of Chaldea. Of the land where women dwell without company of men. Of the knowledge and virtues of the very diamond', *The Travels of Sir John Mandeville*, fourteenth century

A TEXTUALISED WORLD

Although, when travelling beyond the borders of Europe, explorers travelled to new places, this did not mean that these places were entirely unknown to them. There existed knowledge of what was beyond the borders of Europe. These knowledges came from what we can call **'imagined geographies'** based on myth and legend – perhaps most famously, the travels of John Mandeville with which this chapter opened, producing a textualised world rather than one based on observation and experience. These imagined geographies were inhabited by imagined others, people who were very different from Europeans. Indeed, some have argued that it was the existence of these others beyond the borders of Europe that helped to define Europe itself. Edward Said famously developed this idea into his thesis of *Orientalism* to argue that the 'rest' of the world was necessary in order to define Europe. These imagined geographies described the world to people, and explained their place within it, and were thus very significant in shaping how people responded to the world. Although imagined, these geographies had real consequences for people's actions: they were very important to people's understanding of what they saw and experienced in their travels. In this chapter we will see how European understandings and images of the rest of the world emerged.

IMAGINED OTHERS

In the Middle Ages, tales and myths of what European travellers would find beyond known boundaries were common. These imagined others were regarded as monstrous because they were seen as being a transformation of the Europeans in one way or another. Friedman has argued in *The Monstrous Races in Medieval Art and Thought* (1981) that medieval people took 'known' geographic locations and filled them with folk knowledge. These locations were usually just off the map of the experienced world (see Figure 1.1). These were particularly popularised by the writings of Pliny and Mandeville, who described monstrous races just outside of known geographic space.

The chief distinction of these races of people from Europeans then lay in geography. Those who were not European lay 'outside' in faraway, semi-mythological places like India, Ethiopia and Cathay, places whose names evoked mystery and exoticism for Europeans at the time. They were places that Europeans had heard of, but had neither visited nor fully understood.

Medieval scholars listed around 50 peoples. Many of them were frighteningly monstrous but not all. Some were just different in appearance and social practice from those who looked at them. Friedman argues that this difference emerges from a shift in one aspect of their being. Many were visual, for instance:

- *Transformation of body*: these people had huge ears, with their faces on their chest, and were giants or pigmies. For example, the *Amyctyrae* had a protruding lower lip which could be used like an umbrella to protect themselves from the sun. The *Blemmyae*, from the deserts of Libya, were men with faces on their chests.
- *Transformation of Gender*: these tales talked of hairy women, Amazons and androgenes.

Figure 1.1 Medieval map depicting 'monstrous races' around the margins of known geographical space

Other groups of people were identified not by visual differences, but by their difference from European ways of doing things, for example:

- *Transformation of life cycle*. These people were said to rear children just once or to conceive at five years of age.
- *Transformation of social*. Such peoples may look 'normal' but had transformed social practices, such as the wife-givers who were reportedly an amiable race

who gave their wives to any travellers who stopped among them. Other common variants involved cannibals.

- *Transformation of needs.* Here, peoples had entirely different ways of existing. For example, the *Astomi,* apple-smellers, who lived near the headwaters of the Ganges, were said neither to eat nor drink but existed by smelling roots, flowers and fruits, especially apples. It was claimed that they would die if they smelt a bad odour.

It is clear to see how these peoples worked out as Europe's 'other'! The transformation of their physical or social life produced monstrous races as a transformation of the normal order of things, so linking the otherness of non-Europeans to European identity based on normality. Their main characteristic was their **difference** from Europeans. Europeans were always seen as the reference point, Europeans always represented what was right and normal. There were also less bizarrely different peoples. For example, Ethiopians – black men in the mountains of Africa – were understood to have been burnt black because of their close proximity to the sun. Clear empirical evidence (the fact that Europeans could see people with dark skin) seemed to prove the truth of this worldview to the Europeans. The mix of the believable with the incredible made the incredible seem more believable.

The other dominant pre-modern view of Europe's place in the world similarly reinforced a geography of difference. The ancient Greek philosopher Aristotle posited a spherical Earth. This was a theoretical belief rather than something based around experience or empirical evidence. The ancient Greeks had not travelled around the world but they believed the sphere to be the most perfect shape, and thus believed the Earth must be spherical. Aristotle argued that the Earth was split into a number of zones (see Figure 1.2). Greece lay in the 'temperate zone' in the northern hemisphere. To the north, Aristotle posited the existence of a 'frigid zone', and to the south, a 'torrid zone' around the equator. The southern hemisphere, he believed, would be a perfect reflection of the north.

The ancient Greeks believed that life was concentrated in the two temperate zones; the frigid and torrid zones were uninhabited because of the extremes of cold and hot that each place faced. There appeared to be some evidence for these beliefs. From African travellers, the ancient Greeks knew of the existence of desert to the south which seemed to prove the increasing heat towards the equator, and, once again, the travellers themselves, with their skins apparently burnt black by the sun, were further 'evidence' for Aristotle's cosmology.

Cosmologies such as these had a hold on the popular imagination for centuries, long after science had developed new understandings. History records show that in the early fifteenth century Iberian explorers began to seek a

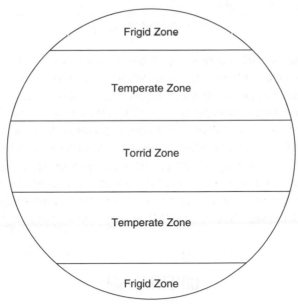

Figure 1.2   Aristotle's cosmology

route around Africa, but were often made to turn back at Cape Bajador where the sea was particularly agitated. While the scientifically educated captains knew this to be a tidal phenomenon, the uneducated crews feared this was the effects of the heat of the torrid zone boiling the water.

It would seem reasonable to assume that with initial travel these monstrous races would disappear. But this did not happen straightforwardly and a belief in the existence of monstrous races persisted. This may have been for two reasons. The first reason was that Europeans had a psychological need to mark their borders with the unknown and different because the demarcation of difference is key to identity (see the box on the following page – we will explore this further later in the chapter). Second, the stories of monstrous races persisted because they did in fact exist! Practices of lip stretching and yoga could seem like distorted bodies to the first travellers; warriors' use of colourful shields might look – from a distance – like faces on their chests; and non-European languages could sound very alien to travellers. Thus, ironically, initial travel may have *reinforced* these mythologies. However, once Europeans started to travel more extensively these stories became less convincing. Gradually there was a greater move towards observation and so the location of the monstrous races changed to being always just beyond the horizon of the known world. However, knowledge of the rest of the world has never escaped textual conventions. We do not see the world innocent of the cultural baggage of our upbringing. It is this idea that lies at the heart of the most important explanation of the way in which the west views the rest – Edward Said's *Orientalism*.

ORIENTALISM

Ship me somewhere east of Suez
Where the best is like the worst
Where there ar'n't no Ten Commandments an' a man can raise a thirst

Rudyard Kipling, Mandalay

Edward Said's 1978 book, *Orientalism*, has probably been one of the most influential texts in the social sciences and arts in the latter part of the twentieth century. Orientalism is conventionally understood to be the scholarly study of the languages and traditions of the Middle East. However, Said argues that Orientalism is not so innocent a form of knowledge as this. Instead, he redefines Orientalism as the ubiquity of a sense of the division of the world into two spheres in aesthetic production, popular culture, and scholarly, sociological, and historical texts. In other words, he is suggesting that the concept of difference between east and west is a geopolitical difference which is written up throughout the texts of western culture whether through travel writing, political texts, paintings, or in academic discussions. To Said, any or all of the cultures of northern Africa, east to southeast Asia and the South Seas could be encompassed by the western geographical imagination into a singular 'Orient'.

For Said, Orientalism is an imaginative geography for two reasons. First, Europeans projected a *single* culture into the space of the 'Orient' that was at odds with the diversity of peoples, cultures and environments contained within the space of the Orient, and second, this space was defined by texts and not by people from the Orient itself. These texts *preceded* experience, so empirical evidence was included but was fitted into the categories that were already constructed. Travellers saw what they expected to see. For Said, this is particularly important because of the link between this imaginative geography and European power. This imaginative geography was made manifest over space as it was built into colonial policy, into the institutions of governance, and more recently, into the practices of aid and development. The imaginative geography of Orientalism shaped the real geographies practised in the space of the Orient.

IMAGINATIVE GEOGRAPHY – SAID'S *ORIENTALISM*

... Orientalism is not a mere political subject matter or field that is reflected passively by culture, scholarship, or institutions; nor is it a large and diffuse collection of texts about the Orient; nor is it representative and expressive of some nefarious 'Western' imperialist plot to hold down the 'Oriental' world. It is rather a *distribution* of geopolitical awareness into aesthetic, scholarly, economic, sociological, historical, and philological texts; it is an *elaboration* not only of a basic geographical distinction (the world is made up of two unequal halves, Orient and Occident) but also of a whole series of 'interests' which, by such means as scholarly discovery, philological reconstruction, psychological analysis, landscape and sociological description, it not only creates but also maintains; it *is*, rather than expresses, a certain *will* or *intention* to understand, in some cases to control, manipulate, even to incorporate, what is a manifestly different (or alternative and novel) world; it is, above all, a discourse that is by no means in a direct, corresponding relationship with political power in the raw, but rather is produced and exists in an uneven exchange with various kinds of power, shaped to a degree by the exchange with power political (as with a colonial or imperial establishment), power intellectual (as with reigning sciences like comparative linguistics or anatomy, or any of the modern policy sciences), power cultural (as with orthodoxies and canons of taste, texts, values), power moral (as with ideas about what 'we' do and what 'they' cannot do or understand as 'we' do). Indeed, my real argument is that Orientalism is – and does not simply represent – a considerable dimension of modern political-intellectual culture, and as such has less to do with the Orient than it does with 'our' world.

[...]

Our initial description of Orientalism as a learned field now acquires a new concreteness. A field is often an enclosed space. The idea of representation is a theatrical one: the Orient is the stage on which the whole East is confined. On this stage will appear figures whose role it is to represent the larger whole from which they emanate. The Orient then seems to be, not an unlimited extension beyond the familiar European world, but rather a closed field, a theatrical stage affixed to Europe. An Orientalist is but a particular specialist in knowledge for which Europe at large is responsible, in the way that an audience is historically and culturally responsible for (and responsive to) dramas technically put together by the dramatist. In the depths of this Oriental stage stands a prodigious cultural repertoire whose individual items evoke a fabulously rich world: the Sphinx, Cleopatra, Eden, Troy, Sodom and Gomorrah, Astarte, Isis and Osiris, Sheba, Babylon, the Genii, the Magi, Nineveh, Prester

(Cont'd)

John, Mahomet, and dozens more; settings in some cases names only, half-imagined, half-known; monsters, devils, heroes; terrors, pleasures, desires. The European imagination was nourished extensively from this repertoire: between the Middle Ages and the eighteenth century such major authors as Ariosto, Milton, Marlowe, Tasso, Shakespeare, Cervantes, and the authors of the *Chanson de Roland* and the *Poema del Cid* drew on the Orient's riches for their productions, in ways that sharpened the outlines of imagery, ideas, and figures populating it. In addition, a great deal of what was considered learned Orientalist scholarship in Europe pressed ideological myths into service, even as knowledge seemed genuinely to be advancing.

Edward Said, *Orientalism*, (1978), pp. 12 and 63

It is important to realise, however, that Said did not consider that Europeans had simply made the Orient up. As he explained, '[o]ne ought never to assume that the structure of Orientalism is nothing more than a structure of lies or of myths which, were the truth about them to be told, would simply blow away' (Said, 1978: 6). He did not oppose the Orientalists' images of the world outside of Europe to reality, but instead understood that they were constitutive of reality because of the way in which knowledge and power were related.

Said developed this idea of the interrelationship between power and descriptions of places from the work of French philosopher Michel **Foucault**. For Foucault power and knowledge are always and everywhere intertwined. He used the term **power/knowledge** not to suggest that power equalled knowledge, but to emphasise the fact that power and knowledge are always and everywhere inseparable. This challenged conventional accounts which suggested that knowledge was repressed by power. In the case of Orientalism, power emerged through institutions and practices used to name and describe the Orient. Those resident in the space of the Orient were not allowed to speak for themselves. They were always described by others, and characterised by others. There is then a power of naming. European taxonomies – the ordering and making understandable of the new world they were exploring – simplified the Orient and, by making it known to Europeans, made it possible for them to control it. The best example of this is the use of maps. Europeans drew maps of new lands with boundaries inscribed to identify territories claimed by different nations. The names given to places by indigenous people were ignored, their claims to ownership or rights of access were similarly discarded, and instead European words and meanings were written onto the maps. Once these European maps had been created and accepted, they started to influence the nature of the actual space they represented. Places took on their European names, reflecting European ownership.

Orientalism was made up of a series of discourses that explained the nature of the Orient and Occident, and the relationship between these two geographical areas. Said argued that is impossible for people to understand the world except through discourse. He is not suggesting that the world is made up only of our imaginations of it, but that we cannot access the real world except through the cultural structurings of discourse (see the box on discourse below).

These discourses were based around a series of binary pairs. At the heart of this, the imaginative geography of Orientalism was a binary geography of the Occident (west) and the Orient (east). However, in western thought, binaries are never different but equal; there is always a hierarchy of values. Thus, Said (1978: 72) insists that Orientals 'are always symmetrical to, and yet diametrically inferior to, a European equivalent, which is sometimes specified, sometimes not'. Taxonomies of difference in the history of western thought have not allowed the existence of 'different but equal'. Western knowledge always imposes a value on the binaries which privileges one term over the other. Sometimes the Occidental value is identified and the Orient is shown to deviate from it; at other times, the Occidental value is universalised, it is 'the' way of doing something and the Oriental equivalent is simply seen as wrong.

DISCOURSE

Discourses define the parameters of what can be known and understood at any point in history and in any place. They can be thought of as a lens through which people interpret the world, which is not unchanging but is temporarily and spatially specific. Discourses do not simply structure knowledge but also what is included as knowledge, such as what are the reasonable questions to ask. For instance, in pre-modern times, religious and mystic discourses dominated understanding. In order to understand an event, people would turn to the Bible or other religious texts, and would look for evidence of God's hand in the world. Scientific discourse sees truth not via faith in God, but in rigorous scientific practice. Scientists look to texts written by other scientists and the laws generated by previous research. Not only do these discourses have different explanations of how the world works, they also look to very different parts of it to justify their beliefs.

There are a number of different themes – or discourses – through which the Orient is marked as being different from Europe (see Figure 1.3).

1   *Development and time.* This discourse could be articulated in a number of ways. Sometimes the Orient was represented as backward while Europe was developed. Alternatively, the Orient was seen as unchanging while Europe was dynamic, as evidenced through the Enlightenment, the drive of mercantile capitalism, or the Industrial Revolution. Some versions of this discourse insisted that cultures were in different stages of development. This discourse recognised

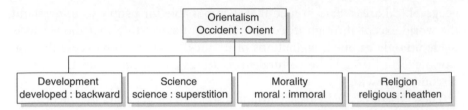

Figure 1.3 Orientalism as a form of representation made up by a number of discourses

that the Egyptians and Chinese had great societies before Europe had developed but that these civilisations were now seen to be in decline. Asia and North Africa were represented as old, decrepit, decaying civilisations. Europe was dynamic yet mature, but sub-Saharan Africa was seen as undeveloped and child-like. This version legitimated European intervention, as it was argued that Europe had come to maturity just as old civilisations like Egypt and China were in terminal decline, so that is was the duty of Europeans to rule the 'immature' peoples in Africa because they were not sufficiently mature to govern themselves.

2   *Morality*. In the discourse of Orientalism, the Orient was immoral and it was the 'white man's burden', as Rudyard Kipling famously put it, to improve the Orient's morals. The discourse of morality was invoked in a number of ways. Moral discourses were used a good deal in assessments of other cultures, of religious practices, even in terms of order and hygiene, which were regarded as expressions of morality. It was sometimes also expressed through sexuality, with the Orient often seen as a place of unrestrained sexuality. This was particularly important during the Victorian era in Britain, when there was considerable sexual repression at home. Oriental women were seen as sexually available whereas men were either seen as hypermasculine with a kind of animal sexuality, or were emasculated, impotent in comparison to the power of western culture. Morality was also encoded through discussions of laziness. Orientals were not considered to be so productive as westerners and travellers often noted that during the day they saw native peoples 'lazing around' rather than working. This was especially important in the moral sense of the Protestant work ethic, where there was value given to hard work with laziness regarded as immoral.

3   *Rationality*. Orientals were seen as irrational, not accepting of European science, and instead turning to animistic beliefs and magic. This notion was particularly important and cross-cuts many other forms of Orientalism, particularly in the eighteenth century onwards where notions of science and reason came to dominate European knowledge, apparently differentiating European views from the 'backward' views of people in other parts of the world.

4   *Religion*. Orientalism did not accept Hinduism, Islam and other non-Christian religions as true religions, and instead saw them as myths or beliefs. Thus, Europeans believed that Orientals were not religious and should be converted to Christianity.

5   *Science.* This seemed to provide 'proof' of western superiority European science had allowed people to conquer nature (in that the landscape was made productive, minerals were mined from the earth, and science offered explanations of how nature worked), time and space (in the power to travel distances relatively quickly using trains, ships and air travel), and the body (in that medicine was providing power over illness). Africans and others were seen to be living with nature. Indeed the fact that they had not created cities and settled cultivation meant to Europeans that natives were unable to exploit natural resources and transform nature: they should thus be seen as part of nature rather than separate from it. Neither did natives understand science, and thus were unable to control their environment. Furthermore, the apparent ease with which European diseases killed many indigenous peoples visited through exploration was 'proof' of Europeans' superior knowledge of the body. That a relatively small number of Europeans were able to take territory into colonial possession seemed to demonstrate the power of European technology and weapons and the superiority of their knowledge. Thus, this 'proof' could be summed up as follows:

- Medicine conquered illness.
- Travel conquered time and space.
- Mining extracted resources from nature.
- Morality controlled natural bodily desires.

More will be said about science in the next chapter, however it is important to note here that scientific discourses were often in conflict with religious discourses. Whereas religious discourses privileged Europeans because of their Christianity, the only true religion, scientific discourses criticised Oriental peoples for their continued reliance on religious beliefs rather than science. Said argues that while the individual discourses comprising Orientalism might be contradictory, the overall structure of the geographical imagination – of the Orient being entirely different from the Occident – remained the same, and therefore reinforced this binary geography of east and west.

6   *Race.* This was not a component of Orientalism until the nineteenth century, when the 'scientific' category of race was used to explain European domination. Through this discourse measurable biological facts (such as head shape or brain size) were used to explain western superiority. We will consider this in more depth in the next chapter.

NATURALISING DIFFERENCE

Even nature was seen as being fundamentally different in the Orient. Consider the following excerpt from Henry Morton Stanley's (1878) account of his travels in Africa, *Through the Dark Continent*:

(Cont'd)

> On the whole, Nature has flung a robe of verdure of the most fervid tints. She has bidden the mountains loose their streamlets, has commanded the hills and ridges to bloom, filled the valleys with vegetation breathing perfume, for the rocks she has woven garlands of creepers, and the stems of trees she has draped with moss; and sterility she has banished from her domain.
>
> Yet Nature has not produced a soft, velvety, smiling England in the midst of Africa. Far from it. She is here too robust and prolific. Her grasses are coarse, and wound like knives and needles; her reeds are tough and tall as bamboos; her creepers and convolvuli are of cable thickness and length; her thorns are hooks of steel; her trees shoot up to a height of a hundred feet. We find no pleasure in straying in search of wild flowers, and game is left undisturbed, because of the difficulty of moving about, for once the main path is left we find ourselves over head amongst thick, tough, unyielding, lacerating grass.

ORIENTALIST ART

One of the main ways that the Orient was made available to Europeans was through art. The ways in which the lands and peoples beyond Europe were represented in painting are instructive of the ways in which the west viewed the rest. Paintings are interesting because of their broad appeal and the fact that at the time, for the majority of Europeans, paintings were the only insight they had into the Orient. Such paintings often presented incredible detail, convincing viewers of their authenticity through the 'reality effect' of lifelike details.

Much of Said's analysis was based upon the written accounts of travellers, academics and politicians. However, we can see a number of the discourses that Said identified encapsulated not only in written descriptions of places, but also in visual representations of them. Indeed, this was one of the central arguments in Said's thesis – that Orientalism cuts across different forms of knowledge, whether textual or visual, academic or popular culture. It is through the resonances between the different types of cultural production (the fact that the message about this geographical difference could be found throughout culture) that Orientalism has developed its influence.

Writing about his painting, *The Fanatics of Tangier*, for the brochure for the 1838 salon, Delacroix stated 'their enthusiasm excited by prayers and wild cries, they enter into a veritable state of intoxication, and, spreading through the streets, perform a thousand contortions, and even dangerous acts'.

He claimed they reached a state of ecstasy which allowed them to walk on red-hot coals, eat scorpions, lick red-hot irons and walk on sword blades, all apparently without noticing their injuries. This image of impassioned religion, steeped in mysticism and secret rites, stood in stark contrast to the restrained and orderly practices of religion in Europe of the time.

Figure 1.4 *The Fanatics of Tangier*, Delacroix, 1838

Figure 1.5 *Dance of the Almeh*, Gerome, 1863

Figure 1.6 *Gateway to the Great Temple at Balbec*, Roberts, 1841

Many Orientalist pictures represented women revelling in the pleasure of wild and released sensuality which would have been impossible to depict in respectable European women at the time. Paintings focused on the erotic, on excess, and male fantasies played out in sites of languid opulence (see Figure 1.5).

Other paintings focused on historical architecture, perhaps most famously in David Robert's images of Egypt (see above). Such paintings overwhelmed the viewer with ruined greatness and an implied criticism of the local people for neglecting their own monuments so that architecture falls into decay. The images within such paintings also reflect the decaying civilisations themselves. Some depict European archealogists restoring the great buildings, suggesting that only European knowledge can recognise the value of the past achievements of these great civilisations and what is thus worth preserving.

ORIENTALISM IN THE PRESENT

However, while it is easy to look back at the past distortions of Orientalism, Said's work insists upon the *continuity* of Orientalism into the present rather than the complete break that previous accounts had suggested. Orientalism is still with us but in a slightly different form, as there is more differentiation between parts of the world (as we shall see in later chapters). Also, importantly, the west is no longer just Europe, now the United States of America has become more influential in the production of dominant images of the rest of the world, particularly through the power of Hollywood. Think about the nature of characters from 'the Orient' in blockbuster movies – can you think of any films where they are the lead heroic characters? If not, what kind of roles do they play (you could look at the *Indiana Jones* series, the *Mummy* films, *True Lies*)?

Postcolonial theorists have also a lot to contribute to discussions about George W. Bush's 'War on Terror'. While none have shown anything but horror at the acts of Al Qa'ida on September 11th 2001, many have been critical of the nature of the response, which has, once again, created a binary imagined geography that has divided the world into the west and the 'axis of evil' to the east.

One of the most influential accounts of the relationship between the west and Islam has been Samuel Huntington's (1993) 'The clash of civilisations' thesis, which (in)famously argued that contemporary international relations would reconfigure around cultural conflict, most significantly between the secular-Christian west and the Islamic east, which were seen to be inherently incompatible. The continuation of Orientalist geopolitics is clear in Huntington's thesis. The events of September 11th seemed to prove the theory, despite various voices, including Said and Huntington himself, which insisted that this was the action of a small group of extremists rather than being representative of Islam more generally. Nevertheless, as Gregory (2004) has highlighted, it is an argument that has been used frequently in media explanations of international relations (see the box below). We will return to contemporary expressions of Orientalism in later chapters.

ORIENTALISM AND THE 'CLASH OF CIVILISATIONS'

Consider the extracts below from an article about the attack on the World Trade Center in New York in September 2001. Think about the language used (especially the discourse of time), and the extent to which Orientalist themes are drawn upon to characterise a distinct geography of 'us' and 'them'.

In This War of Civilisations, the West Will Prevail

(Sir John Keegan, Defence Editor, *The Daily Telegraph*, UK, 08/10/2001, http://www.tele-graph. co.uk/ opinion/main.jhtml?xml=/opinion/2001/10/08/do01.xml, retrieved 13/11/06)

Striking quickly, as well as hard, may be a quality of this war deliberately chosen, and with good reason. A harsh, instantaneous attack may be the response most likely to impress the Islamic mind. Surprise has traditionally been a favoured Islamic military method. The use of overwhelming force is, however, alien to the Islamic military tradition. The combination of the two is certainly designed to unsettle America's current enemy and probably will.

Samuel Huntington, the Harvard political scientist, outlined in a famous article written in the aftermath of the Cold War his vision of the next stage hostilities would take. Rejecting the vision of a New World Order, proposed by President Bush senior, he insisted that mankind had not rid itself of the incubus of violence, but argued that it would take the form of conflict between cultures, in particular between the liberal, secular culture of the West and the religious culture of Islam. Huntington's 'clash of civilisations' was widely discussed, though it was not taken seriously by some. Since September 11 it has been taken very seriously indeed.

[. . .]

The Oriental tradition, however, had not been eliminated. It reappeared in a variety of guises, particularly in the tactics of evasion and retreat practised by the Vietcong against the United States in the Vietnam war. On September 11, 2001 it returned in an absolutely traditional form. Arabs, appearing suddenly out of empty space like their desert raider ancestors, assaulted the heartlands of Western power, in a terrifying surprise raid and did appalling damage.

President Bush in his speech to his nation and to the Western world yesterday, promised a traditional Western response. He warned that there would be 'a relentless accumulation of success'. Relentlessness, as opposed to surprise and sensation, is the Western way of warfare. It is deeply injurious to the Oriental style and rhetoric of war-making. Oriental war-makers, today terrorists, expect ambushes and raids to destabilise their opponents, allowing them to win further victories by horrifying outrages at a later stage. Westerners have learned, by harsh experience, that the proper response is not to take fright but to marshal their forces, to launch massive retaliation and to persist relentlessly until the raiders have either been eliminated or so cowed by the violence inflicted that they relapse into inactivity.

News of the first strikes against Afghanistan indicates that a tested Western response to Islamic aggression is now well under way. It is not a crusade. The crusades were an episode localised in time and place, in the religious contest between Christianity and Islam. This war belongs within the much larger spectrum of a far older conflict between settled, creative productive Westerners and predatory, destructive Orientals.

CRITIQUE OF *ORIENTALISM*

Because of its influence, *Orientalism* has generated a great deal of critical discussion. There are a large number of papers and edited collections discussing Said's work, but there are four particularly important issues.

1   Occidentalism. Just as he critiqued Orientalists of reducing a vast and differentiated area to the Orient, so Said reduces all of Europe (and later also North America) to the Occident. Clearly there are differences within the west. For instance, what differences might there be in the geographical imagination of those countries that did not have colonies? What of internal colonies and groups of others *within* Europe (such as Northern Ireland within the UK, groups of East European Romany travelling people, and those from European empires now living within the west)? Furthermore, there are traditions of 'Occidentalism', representations of Europe and its culture from the non-Western world. These are significant issues, but it is important to remember that although the structures of representation are similar, there is one big difference between Orientalism and Occidentalism which is power, i.e. that the west had, and continues to have, the most powerful voice in representing the west and the rest throughout the world. Historically the influence of Orientalist representations of the world has been much greater than that of Occidentalist accounts, a point we will return to at various places in the book.

2   Historical difference. Said focuses on continuities to the detriment of historical change. While we can trace the continued existence of themes from Orientalism into contemporary culture (and this is in fact something we will do throughout this book), clearly some things are different today – we can see this in the way we view the images of the rest of the world which used to be taken for granted – and if Said's theories are correct, there is no way for accounting for these changes.

3   Gender. Said has been critiqued for an implicit gendering of the Orient as female. Many of the images he used are highly gendered (think of the image in Figure 1.5). Because this is not reflected upon, Said effectively reinforces the patriarchal idea that it is men who are active and capable, and women are passive and unable to represent themselves. As we shall see in Chapter 2, feminists have argued that western women travellers to the Orient produced very different accounts because of the power relations they experienced at home. Their positionality challenged the neat binaries that Said's work depends upon.

4   Retextualisation. Said talks of a textualised Orient, but in his work he does not detextualise it but retextualises it. Following from the Foucaultian literature he draws on, no-one can provide a true representation of reality, all is constructed through discourse. Now the Orientalists' texts are replaced by Said's text. While the Orientalist texts aimed to show that the Orient was backward, this has been replaced by Said's aim of demonstrating the political nature of the Orientalists. The values (whether the Orient or the Occident contains the problem) have changed but the structure of representation (there is a geographical space of Orient and of Occident) remains the same. We still do not know what 'they' think of themselves, as the voices of Oriental people are not included in the

book. Moreover, if all versions of the Orient are textual creations, how are we to argue that today's version, or Said's version, is any better or worse than those presented in the nineteenth century by European colonisers and earlier travellers? This is a particularly important critique for us as geographers as one of the tasks that we have traditionally had is to write about different parts of the world. As we shall see, geography as a discipline was very much part of the European colonial enterprise and was caught up in Orientalism. Are we still Orientalists if we seek to write geographies of the Middle East or Asia?

Further reading

On early European views of the rest of the world

Baudet, H. (1988) *Paradise on Earth: Some Thoughts on European Images of Non-European Man* (trans. E. Wentholt). Middletown, CT: Wesleyan Press.

Friedman, J. (1981) *The Monstrous Races in Medieval Art and Thought*. Harvard: Harvard University Press. (See especially the Introduction and Chapters 1 and 2.)

The classic text on the way the rest of the world has been represented by the west

Said, E. (1978) *Orientalism*. New York: Vintage.

For further discussion of Said's work

Ahmad, A. (1992) 'Orientalism and after', in A. Ahmad, *In Theory*. London: Verso. (Reprinted in P. Williams and L. Chrisman (eds) (1994) *Colonial Discourse and Post-colonial Theory*. New York: Columbia. pp. 162–71.)

Lewis, R. (1996) *Gendering Orientalism: Race, Femininity and Representation*. London: Routledge.

Porter, D. (1994) 'Orientalism and its problems', in P. Williams and L. Chrisman (eds), *Colonial Discourse and Post-colonial Theory*. New York: Columbia. pp. 150–61.

Said, E. (1985) 'Orientalism reconsidered', *Race & Class*, 27(2).

Young, R.J.C. (1990) *White Mythologies: Writing, History and the West*. London: Routledge. (See especially Chapter 7.)

Shohat and Stam develop Said's arguments to a world where the dominant representation comes through US media

Shohat, E. and Stam, R. (1994) *Unthinking Eurocentrism*. London: Routledge.

On contemporary expressions of Orientalism in international relations (especially concerning the 'war on terror')

Gregory, D. (2004) *The Colonial Present*. Oxford: Blackwell.

2

KNOWLEDGE AND POWER

This book first arose out of a passage in Borges, out of the laughter that shattered, as I read the passage, all the familiar landmarks of my thought – *our* thought, the thought that bears the stamp of our age and our geography – breaking up all the ordered surfaces and all the planes with which we are accustomed to tame the wild profusion of existing things, and continuing long afterwards to disturb and threaten with collapse our age-old distinction between the Same and the Other. This passage quotes a 'certain Chinese encyclopaedia' in which it is written that 'animals are divided into: (a) belonging to the Emperor, (b) embalmed, (c) tame, (d) suckling pigs, (e) sirens, (f) fabulous, (g) stray dogs, (h) included in the present classification, (i) frenzied, (j) innumerable, (k) drawn with a very fine camelhair brush, (l) *et cetera*, (m) having just broken the water pitcher, (n) that from a long way off look like flies'. In the wonderment of this taxonomy, the thing we apprehend in one great leap, the thing that, by means of the fable, is demonstrated as the exotic charm of another system of thought, is the limitation of our own, the stark impossibility of thinking *that*.

But what is it impossible to think, and what kind of impossibility are we faced with here? Each of these strange categories can be assigned a precise meaning and a demonstrable content; some of them do certainly involve fantastic entities – fabulous animals or sirens – but, precisely because it puts them into categories of their own, the Chinese encyclopaedia localizes their powers of contagion; it distinguishes carefully between the very real animals (those that are frenzied or have just broken the water pitcher) and those that reside solely in the realm of imagination. The possibility of dangerous mixtures has been exorcized, heraldry and fable have been relegated to their own exalted peaks: no conceivable amphibious maidens, no clawed wings, no disgusting, squamous epidermis, none of those polymorphous and demoniacal faces, no creatures breathing fire. The quality of monstrosity here does not affect any real body, nor does it produce modifications of any kind in the bestiary of the imagination; it does not lurk in the depths of any strange power. It would not even be present at all in this classification had it not insinuated itself into the empty space, the interstitial blanks *separating* all these entities from one another. It is not the 'fabulous' animals that are impossible, since they are designated as such, but the narrowness of the distance separating them from (and juxtaposing them to) the stray dogs, or the animals that from a long way off look like flies. What transgresses the boundaries of all imagination, of all possible thought, is simply that alphabetical series (a, b, c, d) which links each of those categories to all the others.

Michel Foucault, 'Preface', The Order of Things,
pp. xv–xvi

The opening quotation, the preface from *The Order of Things* (1970), where the author Michel Foucault explains his motivations for writing the book, seems like a bizarre turn of fantasy. As he says, who could possibly believe such things? Foucault's intention in quoting this strange Chinese encyclopaedia was to point out that the role of knowledge is to create orders so that we can make sense of the multitude of information about the world around us. Things are grouped into categories and associations that allow us to navigate our way through the world. These groupings make order out of the chaos of stuff around us. Foucault argues that through history different 'episteme', or ways of knowing or organising the world, have come to dominate.

Foucault argues that the beginning of the nineteenth century witnessed the rise of the modern episteme. People began to classify difference and discontinuity into taxonomies of knowledge which were categorised into modern differentiations such as culture *or* nature, modern *or* traditional, science *or* mythology. With the rise of exploration, there was a real fear of the unknown and a desire to chart and explain new worlds. Formal colonialism generated an added incentive to learn about what had been conquered, whether in order to manage the natives and learn about their ways so as to control them (and levy taxes), or to find out about the resources available that would boost the wealth of a colonising power. Geographers had an important role in this. Indeed, the formation of the modern discipline was dependent upon geographers' roles in the charting of new places and acted as an aid to statecraft in expansion overseas.

In this chapter we will look at the rise of scientific understandings of the world, the institutions through which colonial knowledge was generated and developed, and the role that it played in governance. We will then go on to explore the ways in which this knowledge was communicated to the general population; how the majority of people in Europe found out about the rest of the world.

KNOWING THE WORLD

As we saw in the previous chapter, discourses of difference from the Medieval period to the seventeenth century were based upon religion, appearance and social patterns. Science and technology were not often used as China, India and the Middle East were more advanced than Europe in many technologies and forms of knowledge. In the eighteenth century, the **European Enlightenment** saw a rise in the importance of science and technology as the basis of comparison. Capitalism drove the economies of Europe and scientific knowledge was used to develop new methods of production. Europeans saw their economy and society developing quickly, and viewed themselves as dynamic and vibrant in comparison to what seemed to them as timeless cultures, or even decaying civilisations elsewhere.

Time and space were central. European modernity – especially the rise of industrial production – meant that time and organisation were very important (think of the centrality of the organisation of time to the modern world, particularly the significance of timetables and workshifts). Other societies were seen to fail to 'value' time, adhering to natural rhythms or seemingly unable to use time efficiently. Space was similarly valued through the knowledges produced by charts and maps and the centrality of space in the layout of the industrial production line. This mathematisation of space allowed for better management and control, whether in terms of better navigation, or the more efficient and accurate measurement of space for the collection of taxation. Railways opened up continents and helped to link countries together. They also facilitated a more efficient exploitation of resources and the movement of troops around countries. This again seemed to provide evidence of European superiority through technology (and therefore, by extension, European civilisation): with relatively small numbers of troops they could control a country. Europeans triumphed over space, whereas others were trapped in place.

Religion was still important to many as a category, but was nonetheless in decline. This trend was reinforced in the nineteenth century where science and technology, modernity and progress became paramount in the understanding of differences between cultures. For the first time in the nineteenth century came the rise of a racial 'science' which sought to explain the differences between peoples.

COLONIAL KNOWLEDGE: MACHINES
AS THE MEASURES OF MEN

In many places, European travellers found no indigenous written sources and so they presumed an absence of science and philosophical learning. Science was seen as a very important tool which demonstrated human achievement over nature. In theoretical terms, the development of science allowed for the development of the laws of the physics and chemistry, and the ability to conceptualise abstract problems. In practical terms, it facilitated precision, engineering and measurement. In his book *Machines as the Measures of Men* (1989) , Michael Adas demonstrated the central importance of scientific knowledge in definitions of European superiority over others so that 'a society's level of development could be gauged by its technological achievements' (Adas, 1989: 100).

There was great debate about whether non-Europeans could be taught about science, or whether there was a fundamental difference in abilities. It was often concluded that they could be taught to operate technology, but not to comprehend abstract science. What was important about these discourses of science was the fact that they appeared value-neutral and thus beyond reproach. This was not about subjective judgement or insults; European science seemed entirely rational and unbiased, as Adas explains:

> Because nineteenth century Europeans believed that machines, skull size, or ideas about the configuration of the solar system were culturally neutral facts, evaluative criteria based on science and technology appeared to be the least tainted by subjective bias. (1989: 145–6)

With the Enlightenment came ideas about the separation of 'man' from nature. It was believed that resources were put on the Earth for man, and thus he was responsible for their use (and it was 'man' and not 'humans'). There is a clear link here between Europeans' mastery of nature and the domination of other peoples around the globe who still seemed to be part of nature and not separate from it, namely 'savages' had done little to reshape their environment. Perhaps today this judgement seems a little odd to us. We may hold that the idea of living in a sustainable way with the environment, and making little impact, is a good thing. However, in the eighteenth and nineteenth centuries nature was still something that was seen to be powerful and threatening to humans, and so there was a very strong view that nature had to be tamed. This fear of nature and the native people who lived with it provided an important spur to action. There was a clear sense that if colonisers knew what lay within the country they sought to rule, and if they could truly know and understand it, it would be easier to govern and control.

Despite its ascendance as the predominant form of differentiation between societies, science was not always used in isolation. There were various interdependencies between scientific and religious forms of description, despite their apparent contradictions. For instance, the engineering triumph of the railways was also seen as being good for religious conversion. The explorer David Livingstone, 'who saw himself as a "cog in God's machinery", regarded railroads and telegraphs as important instruments for breaking down barriers

Figure 2.1 This picture is of the geographer and explorer Alexander von Humboldt with his travelling companion Aimé Bonpland. Look at the way in which they have been represented as men of science, surrounded by specimens and equipment to take measurements and readings (and compare this with the simplicity of the hut behind them)

to Christian conversion' (Ibid., p. 206). One nineteenth century European commentator suggested that 'Thirty miles an hour is fatal to the slow deities of paganism' and called it 'pilgrimage done by steam' (Edwin Arnold, 1882, quoted in Adas, 1989: 226).

ACADEMIA AND COLONIALISM

The eighteenth century saw a significant rise in academic societies willing to assist in the colonial endeavour. In 1859 the Anthropological Society of Paris was established and this in turn became a model for others.

In the mid-nineteenth century the first geographical societies emerged in Europe. In Britain, the Royal Geographical Society was established in 1830. Although this concentrated upon the collection of academic information through the sponsorship of exploration, it too had a close relationship with empire. Table 2.1 shows this clearly. Stoddart (1986) listed the occupation of 304 of the original 460 members and this showed that the influence of the

Table 2.1 Founding membership of RGS by profession (from Stoddart, 1986: 60)

Dukes	3
Earls	9
Other peers	24
Baronets, knights	38
Naval officers	32
Army officers	55
Fellows of the Royal Society	124
Fellows of the Geological Society	19
Total	460

aristocratic and military classes to be disproportionate to their numbers in society as a whole.

For effective rule to be maintained, the tenet was 'know your natives'. Knowledge was the charter for domination. Military pacification was followed by academics and royal commissions to understand sources of resistance and counter resistance. Knowledge was used to produce a skilled and pliant labour force; to reduce resistance; to establish forms of governance and taxation; to maximise resource usage.

In her work on the French colonisation of Egypt, Anne Godlewska (1994) shows the importance of the discipline of Geography and the support it received from Napoleon. Geographers could chart what had been conquered, could identify natural resources, could highlight potential places of native resistance and thus how best to deploy troops in a new land. Geographical knowledge was key to the taking and subsequent ruling of Egypt and other colonial possessions. At the same time as offering this very practical support of empire, geographical knowledge also apparently demonstrated the superiority of the French over the Egyptians due to the ease with which the country was taken and then brought under French control. The maps of France's newly acquired territories (again produced by geographers) were a potent reminder to the French population of the value of being part of such a great country, and so Godlewska argued that at the same time as facilitating conquest abroad, geographical knowledge also helped governance at home. No wonder Napoleon was so supportive of this new discipline!

Geographical theories of the time also assisted colonialism in less direct ways. The end of the nineteenth century and the early part of the twentieth witnessed the rise of '**environmental determinism**', an approach that regarded human beings or human society as being the product of the environment within which they lived. Many geographers put this argument forward, but best known among them was Ellsworth Huntington, and especially his book from 1907, *The Pulse of Asia*, which recorded his travels within the continent and ended with his argument about the correlation between climate types and civilisation, an argument developed more fully in *Civilization and Climate* in 1915. Huntington produced maps of the world which showed the potential places where civilisation could emerge. These were centred around the cool, invigorating climes of northwestern Europe. Further north, he argued, and it was considered too cold for effective civilisation to emerge; further south, and people were rendered too languid due to the heat to go about building civilisation.

KNOWING THE NATIVE BODY

It was not only geographical societies that emerged and flourished at this time. The discipline of anthropology was established at the same time as geography and for many of the same imperialistic reasons. Much of the work of this discipline was concerned with 'physical anthropology', which measured body parts as a 'scientific' way to classify races. By the end of the nineteenth century, the use of photography further assisted this science of racism. Thousands of images of natives were used to create racial archetypes. Natives' bodies were photographed against measuring sticks in front of a grid so as to allow for detailed calculations of all the dimensions of a body. This form of science led to various attempts to explain the differences between 'races' through rational, scientific explanation. It is now clear how problematic this 'science' was and much of it was simply thinly disguised racism. And yet such explanations held great sway and were generally considered entirely respectable at the time.

Racism relates human behaviour and character to the race (phenotype) to which an individual or group belongs. One very important focus for racist scientific measures of difference was the body. Racist ideology usually involved an aesthetic appraisal of physical features and an elaborate classification of traits of mind and personality linked to physical features. This relied upon an implicit mind-body unity within which the shape of the body bore witness to the soul. In the late eighteenth century racial hierarchies were constructed with whites (especially north-west Europeans) at the top and Jews and blacks in lower places, and some went further to link this to the new science of evolution by including apes at the bottom. These hierarchies were established through the measurement of skull capacity, brain size, face shape, and various other aspects of the body. It was argued that black physical traits – dark skin, coarse hair, thick lips – were outward signs of 'inner cognitive defect'. This 'science' was cross-cut by aesthetics. Cuvier (1827–1835), in the 16-volume *The Animal Kingdom*, compared black people to monkeys and stated that they would forever remain in barbarism. In contrast, he argued that Europeans are distinguished by:

> The Caucasian, to which we belong, is distinguished by the beauty of the oval formed by his head, varying in complexion and the colour of the hair. To this variety, the most highly civilized nations, and those which have generally held all others in subjection, are indebted for their origin.
>
> (Curvier, 1832: 50)

The initiators of racism were European intellectuals: clergymen, physicians, professors, and philosophers. In the nineteenth century, science was pressed into the service of racism to prove the inferiority of blacks, all colonised people outside of Europe and Jews in Europe. Such views seemed to gain scientific legitimacy from Darwin's influential *The Origin of Species by Means of Natural Selection*, published in 1859. Darwin introduced the notion of a 'struggle for life', in which some species would always be more successful than others in adapting to local natural-environmental conditions. Although Darwin's ideas revolved around changes that would occur over many generations, and from random fluctuations in genetic adaptation, the idea of the 'survival of the fittest' inherent in the struggle-for-life argument was focused on by environmental determinists. These 'social Darwinists' viewed the world as being split into winners and losers, with the winners being those organisms (or people) most able to adapt to prevailing environmental circumstances and the losers being those who were unable to adapt and fit only for death.

Social Darwinism can also be seen as an apology for the excesses of capitalists and imperialists who were associated with imperial expansion at the time, making political struggle, class warfare, economic competition and rapid change and upheaval seem like a natural state for human life:

> Capitalism in its red in tooth and claw competitive stage provided the social model for a new mode of natural understanding. In turn, natural science provided legitimation for conducting social life in this dog-eat-dog way. (Peet, 1985: 313)

Social Darwinism, understood within the context of environmental determinism, made conflict, violence and domination seem natural, and an inevitable feature of human life. This gave a free rein to capitalists and imperialists to do as they wished, as it would appear that winners and losers were therefore also natural. Social Darwinists argued that in the struggle for existence, some peoples would inevitably be eliminated by superior 'races'. While such extreme views were discredited later in the century for their questionable scientific basis and racist assumptions (as we shall see later), there are elements of such views that still persist within contemporary imagined geographies.

Table 2.2   Binaries at the heart of western reasoning

Mind	Body
Reason	Passion
Culture	Nature
Active	Passive
White	Black
Male	Female
Middle class	Working class
Adult	Child

Cross-cutting all this was the belief that we have come across already that other races were more closely related to nature than whites, that they were more tied to their bodies than their minds. This returns us to the centrality of **binary logic** at the heart of the European Enlightenment thought that we have mentioned already. Enlightenment binary concepts separated the mind and rationality from nature and the body, and this central belief was reinforced by the existence of other structuring binaries running through western thought (see Table 2.2). Other races were considered to be more natural in their instincts, and sometimes even viewed as having animal passions – more given over to the body and helpless in the face of desires that were both physical and sexual. This idea was linked to a perceived lack of control over the senses that was achieved through the development of the mind. Women, the lower classes, and children were also held to be more embodied than educated white men. What links those concepts (to the right of Table 2.2 above) is embodiment, or more correctly, a lack of control over the body by the mind: women were regarded as being closely aligned to natural patterns due to their role as mothers, through the cycle of menstruation and 'womanly' problems such as hysteria and fainting; the working classes were believed to be lacking in intellect and driven by bodily passions rather than higher goals; children were viewed as not yet sufficiently developed to achieve control and reason. All had to be controlled by patriarchal powers – elites, fathers, and imperialists. Once again we see that European men positioned themselves as normal, against which all others were compared and from which all were seen to deviate.

So far, this chapter has discussed formal science and academia. While disciplines such as geography and anthropology were becoming more influential within universities and their knowledge was having greater influence within government and colonial policy, the majority of the population of Europe were not aware of the developments in knowledge that had been made. Said, however, argued that the power of Orientalism lay in its ubiquity, that the division between east and west was (and still is) fundamental to western knowledge in both high culture and popular culture. We will now turn to look at the ways in which the majority of the European population came to know about the rest of the world.

TRAVELLERS' TALES

Not the least interesting part in the study of geographical discovery lies in the insight it gives on into the characters of that special kind of men who devoted the best part of their lives to the exploration of land and sea. In the world of mentality and imagination which I was entering it was they and not the characters of famous fiction who were my first friends. Of some of them I had soon formed for myself an image indissolubly connected with certain parts of the world. For instance, western Sudan, of which I could draw the rivers and principal features from memory even now, means for me an episode in Mungo Park's life.

It means for me the vision of a young, emaciated, fair-haired man, clad simply in a tattered shirt and worn-out breeches, gasping painfully for breath and lying on the ground in the shade of an enormous African tree (species unknown), while from a neighbouring village of grass huts a charitable black-skinned woman is approaching him with a calabash [a gourd] full of pure cold water, a simple draught which, according to himself, seems to have effected a miraculous cure. [...]

...the monuments left by all sorts of empire builders [will not] suppress for me the memory of David Livingstone. The words 'Central Africa' bring before my eyes an old man with a rugged, kind face and a clipped, gray moustache, pacing wearily at the head of a few black followers along the reed-fringed lakes towards the dark native hut on the Congo headwaters in which he died, clinging in his very last hour to his heart's unappeased desire for the sources of the Nile.

Joseph Conrad (1926), 'Geography and Some Explorers' in *Heart of Darkness*

Terra incognita: these words stir the imagination. Through the ages men have been drawn to unknown regions by Siren voices, echoes of which ring in our ears today when on modern maps we see spaces labeled 'unexplored', rivers shown by broken lines, islands marked 'existence doubtful'.

Wright (1947), *President of the Association of American Geographers.*

Travellers' tales have always been popular due to their mix of excitement, romance, and exoticism. Dreams of far-off places often seem to involve child-like and innocent enthusiasm. However, as we have seen, travel and exploration were caught up in the processes of imperialism and colonial expansion that characterised nineteenth century imagined geographies. Explorers were at the forefront in the establishment of colonies; recording, measuring and collecting information about new lands and new peoples. Travellers' and explorers' accounts played an important role in communicating information about new lands and peoples to Europe's populations, in addition to normalising particular representations of colonial rule. In Britain in the nineteenth century the popularity of such accounts was so great that travel books were outsold only by religious books, and travellers' illustrated talks around the country packed in the crowds. These tales constructed the world through narratives of heroism and bold deeds. They drew upon the kinds of discourse already introduced, but made the representations much more accessible to the population at large.

The style of these explorers' accounts was significant in two ways. First, the way that explorers looked at the world, according to Mary-Louise Pratt's influential (1992) critique, could be characterised by their adoption of the 'monarch of all I survey' rhetoric. Pratt has argued that explorers tended to look down upon new scenes to be described – on the one hand, to provide a vantage point from which to get a good view, but, on the other, she argued, this then had more profound impacts on the way in which the explorer described his (and the gendering is deliberate) relationship with the land. Not only does standing at a high point raise an explorer above the landscape and people being described, already implying a relationship of power, this privileged point of view also places the explorer outside of what is being described, and establishes his viewpoint as authoritative. French theorist Michel de Certeau described this perspective as follows:

> His elevation transforms him into a voyeur. It puts him at a distance. It transforms the bewitching world by which one was 'possessed' into a text that lies before one's eyes. It allows one to read it, to be a solar Eye, looking down like a god. The exaltation of a scopic and Gnostic drive; the fiction of knowledge is related to this lust to be a viewpoint and nothing more. (de Certeau, quoted in Ryan, 1996: 6)

The explorer's account from this vantage point, then, is presented as the most accurate and complete (for an example of this, see the box on Henry Morton Stanley on p. 41).

The second important aspect about the nature of explorers' accounts was the structure of the narrative of their travels. The first-person account articulated through an explorer's journal ensured that he was at the centre of the story, creating him as the hero of that story:

> The mythology of exploration insists that the explorer be pitted against the vicissitudes of nature, hounded by inconsiderate indigenes and worn out by hunger in the service of his country. Above all, the explorer is an heroic *individual*: in exploration hagiographies there is rarely mention of the other members of the party ... (Ryan, 1996: 21)

Richard Philips (1997) and others have argued that this exploration was marked by the convergence of nationalism, masculinity and the romance of far-off places. It was a kind of muscular, individualistic, independent masculinity. This was not a bookish intellectual, but instead a figure with great knowledge of survival skills, nature and what has been called the 'sternly practical pursuit' of geography. The narrative excludes women, unless they represent ties to home (wives and mothers who urge them not to go, who try to domesticate them, loved ones and the safety longed for when far away), or prizes or things to protect. This was promoted in literature for adults but could also be seen in children's stories, particularly in the exciting 'Boys' Own' stories, which some have argued were significant in grooming the next generation of colonisers.

It is important to put this into the context of conceptions of sexuality in Europe at the time, especially in Victorian Britain. The rise of bourgeois society and middle class morality, especially in the nineteenth century, brought in prudish ideas of sexuality, particularly those involving repression and control. Within Britain the middle classes worried about working-class practices of sexuality and morality. The middle classes felt that the working classes could not control themselves and they agonised over the negative effects of urban life, especially for women. This in turn paralleled their understanding regarding the sexuality of equally embodied other peoples.

Thus it became necessary to demonstrate the 'proper' practices of morality: in the duty of patriotism; in the moral and physical beauty of athleticism; in the salutary effects of Spartan habits and discipline; in the cultivation of all that was masculine and the expulsion of all that was effeminate, unEnglish and excessively intellectual – what geographer David Stoddart (1986) has celebrated as 'solid hunks of British manhood'.

The narrative of the heroic explorer was that such figures opened up blank spaces on the map for colonialism and imperialism by defeating barbarism and overcoming the challenges of the natural world. They battled with the elements before the oncoming domesticating forces of colonialism:

> Since continued expansion represented a means to achieve or maintain moral, racial, spiritual, and physical supremacy, exploration thus becomes an instrument not only to justify imperial or nationalist political doctrine, but also to embody the supposed collective cultural superiority of the nation. (Riffenburgh, 1993: 2)

Such a narrative reduced stories to the image of the individual spirit versus the wilds, making for an innocent, natural and heroic tale. In actual fact, this was not always the case. When Stanley left Zanzibar to cross Africa in 1874 he took three white assistants, 356 native bearers and labourers, eight tonnes of stores and a 40-foot boat that had to be taken apart to be carried through the jungle.

Explorers popularised their tales and deeds in books and presentations on lecture circuits of the amateur geographical societies which had arisen during the nineteenth century. The Royal Geographic Society (RGS) was one of most fashionable London societies. Their 'Africa nights' to discuss the latest feats of exploration met with 'an immense audience [which] thunders at the gate' (quoted in Driver, 1991: 144). The popularity of these travel accounts was especially significant in the nineteenth century with the rise of the popular press. It was, after all, the most important way in which people learnt about the world.

STANLEY AND HEROISM

Think about the style adopted in the excerpts below from *Through the Dark Continent*, by H.M. Stanley (1878). Consider the claims of authority, heroism and masculinity presented here. In the last excerpt, think about the claims he makes for the objectivity of his viewing of the landscape.

> Unless the traveller in Africa exerts himself to keep his force intact, he cannot hope to perform satisfactory service. If he relaxes his watchfulness, it is instantly taken advantage of by the weak-minded and the indolent.
>
> [...]
>
> What a forbidding aspect had the Dark Unknown which confronted us! I could not comprehend in the least what lay before us. Even the few names which I had heard from the Arabs conveyed no definite impression to my understanding. What were Tata, Meginna, Uregga, Usongora Meno, and such uncouth names to me? They conveyed no idea, and signified no object; they were barren names of either countries, villages, or peoples, involved in darkness, savagery, ignorance, and fable. [...]
>
> The object of the desperate journey is to flash a torch of light across the western half of the Dark Continent. For from Nyangwé east, along the fourth parallel of south latitude, are some 830 geographical miles, discovered, explored, and surveyed, but westward to the Atlantic Ocean along the same latitude are 956 miles — over 900 geographical miles of which are absolutely unknown. [...]
>
> A thousand things may transpire to prevent the accomplishment of our purpose: hunger, disease, and savage hostility may crush us; perhaps, after all, the difficulties may daunt us, but our hopes run high, and our purpose is lofty; then in the name of God let us set on, and as He pleases, so let Him rule our destinies!
>
> [...]
>
> On the 19th a march of five miles through the forest west from Kampunzu brought us to the Lualaba, in south latitude 3° 35', just forty-one geographical miles north of the Arab depot Nyangwé. An afternoon observation for longitude showed the east longitude 25° 49'. The name Lualaba terminates here. I mean to speak of it henceforth as THE LIVINGSTONE.
>
> [...]
>
> Now look at this, the latest chart which Europeans have drawn of this region. It is a blank, perfectly white. [...]

(Cont'd)

I assure you [. . .] this enormous void is about to be filled up. Blank as it is, it has a singular fascination for me. Never has white paper possessed such a charm for me as this has, and I have already mentally peopled it, filled it with the most wonderful pictures of towns, villages, rivers, countries, and tribes – all in the imagination – and I am burning to see whether I am correct or not. *Believe?* I see us gliding down by tower and town, and my mind will not permit a shadow of doubt. Good night, my boy! Good night! and may happy dreams of the sea, and ships, and pleasure, and comfort, and success attend you in your sleep! Tomorrow, my lad, is the day we shall cry – 'Victory or death!'

From my loft eyrie I can see herds upon herds of cattle, and many minute specks, while and black, which can be nothing but flocks of sheep and goats. I can also see pale blue columns of ascending smoke from the fires, and upright thin figures moving about. Secure on my lofty throne, I can view their movements, and laugh at the ferocity of the savage hearts which beat in those thin dark figures ... As little do they know that human eyes survey their forms from the summit of this lake-girt isle as that the eyes of the Supreme in heaven are upon them.

Subverting dominant travel? 'Lady travellers'

A lady explorer? a traveller in skirts?
The notion's a trifle too seraphic:
Let them stay and mind the babies, or hem our ragged shirts;
But they mustn't, can't, and shan't be geographic.

Punch magazine's song to the Royal Geographical Society, 10 June 1893, quoted in Stoddart, 1986: 63)

The myth of the explorer was a tale very much caught up in masculinism. As we can see in the quote above, the prevailing view of the RGS, if not society more generally, was that women were better off at home, not out there in the wilds discovering places.

However, women's exclusion from travellers' tales was not only exercised through their exclusion from professional societies and formal expeditions. The very nature of heroic tales of exploration was masculinist and thus exclusive of women. In these narratives, nature was coded as female and often written using a language of seduction. Furthermore, as we have already seen, Mary-Louise Pratt (1992) has argued that for the heroic male explorer the most common representation of land was from a birds-eye position (like a map). Again, following up on the binaries of western knowledge, the masculine knower is disembodied and distant from the land he is viewing.

But for women, this position is different. In contrast with male discovery rhetoric, seeing violates the norms of conduct for women. In western culture, as we have seen, there is a series of binaries which underpin our structures of

Figure 2.2   Jan van der Straet's *America* (look at this representation of Columbus arriving in America. What are the characteristics of male and female, and of coloniser and colonised, as shown in this image?)

thought. The differences between men and women are reinforced by other pairs in each binary. While men are seen as active agents, women are seen as passive – men look and women are looked at. If this seems like an exaggeration, think about the artistic tradition of painting nudes. If you go into a museum to look at such great works, the vast majority of painters will be men and their subjects will usually be women. The male artist is active (he has painted the picture) while the women are passive, sitting (or more often lying) still while they are represented. Similarly, a male viewer of the picture will be active, doing the looking. The woman in the picture will often be both naked and submissive (either looking away or up at the viewer).

Feminist theorists have suggested that this relationship is much more significant than simply being a convention in art. Thus, a viewer is characterised as being male and that which is being seen is coded as female (think of 'virgin territories' awaiting discovery by male explorers). It is not that women cannot look (at pictures, landscapes or other women), but that the gendered relationship is structured in such a way as to make this difficult, and we see this encompassed in the particular strategies adopted by early female travellers.

For instance, one English explorer of West Africa, Mary Kingsley, depicts herself discovering swamps not by looking down on them or walking around them to demarcate them, but by sloshing zestfully through them. Kingsley was not taking possession of what she saw but instead stole past. Why does this difference exist?

Women were not the first European people to 'discover' places. Audiences were interested in their books and talks because they were the first European *women* to set foot in a new place, and they were of further interest because they were thus doubly out of place. In addition, they had different reasons for such travel. It was impossible for them to take part in scientific expeditions due to their complete exclusion from scientific societies until the dawn of the twentieth century (the RGS did not admit women as fellows until 1913), and rarely did they have the independent means required due to the position of women in European society at the time. When they did write of their experiences, there were general expectations that women would choose to write personalised accounts of their experiences rather than seek to contribute to science. There were also many other barriers to those women who wanted to travel. For instance, Mary Kingsley was a keen naturalist from family of scientists. However, she was the one required to stay at home and look after her parents when they fell ill. It was only after they died and left her money, when she turned 30, that she was able to see for herself what her male relatives had already seen.

MARY KINGSLEY, *TRAVELS IN WEST AFRICA, 1897*

Having been escorted by half of the population for a half mile or so beyond the town, and being then nervous about Fans [local people], from information received, I decided to return to Kangwe by another road, if I could find it. I had not gone far on my quest before I saw another village, and having had enough village work for one day, I made my way quietly up into the forest on the steep hillside overhanging the said village. There was no sort of path up there, and going through a clump of shenja, I slipped, slid, and finally fell plump through the roof of an unprotected hut. What the unfortunate inhabitants were doing, I don't know, but I am pretty sure they were not expecting me to drop in, and a scene of great confusion occurred. My knowledge of Fan dialect then consisted of Kor-kor, so I said that in as fascinating a tone as I could, and explained the rest with three pocket handkerchiefs, a head of tobacco, and a knife which providentially I had stowed in [. . .] my pockets. I also said I'd pay for the damage, and although this important communication had to be made in trade English, they seemed to understand, for when I pointed to the roof and imitated writing out a book for it, the master of the house said 'Um', and then laid hold of an old lady and pointed to her and then to the roof, meaning clearly I had equally damaged both, and that she was equally valuable.

[. . .]

The old male [gorilla] rose to his full height (it struck me at the time this was a matter of ten feet at least, but for scientific purposes allowances must be made for a lady's emotions) and looked straight towards us . . .

[. . .]

About five o'clock I was off ahead and noticed a path which I had been told I should meet with, and, when met with, I must follow. The path was slightly indistinct, but by keeping my eye on it I could see it. Presently I came to a place where it went out, but appeared again on the other side of a clump of underbush fairly distinctly. I made a short cut for it and the next news was that I was in a heap, on a lot of spikes, some fifteen feet or so below ground level, at the bottom of a bag-shaped game pit.

It is at these times you realise the blessings of a good thick skirt. Had I paid heed to the advice of many people in England, who ought to have known better, and did not do it themselves, and adopted masculine garments, I should have been spiked to the bone, and done for. Whereas, save for a good many bruises, here I was with the fullness of my skirt tucked under me, sitting on nine ebony spikes some twelve inches long, in comparative comfort, howling lustily to be hauled out. The Duke came along first, and looked down at me. I said, 'Get a bush-rope, and haul me out.' He grunted and sat down on a log. The Passenger came next, and he looked down. 'You kill?' says he. 'Not much,' say I; 'get a bush-rope and haul me out.' 'No fit', says he, and sat down on the log. Presently, however, Kiva and Wiki came up, and Wiki went and selelcted the one and only bush-rope suitable to haul an English lady, of my exact complexion, age, and size, out of that one particular pit.

[…]

From the deck of the *Niger* I found myself again confronted with my great temptation – the magnificent Mungo Mah Lobeh – the Throne of Thunder [Mount Cameroon]. Now it is none of my business to go up mountains. There's next to no fish on them in West Africa, and precious little rank fetish [Kingsley's study was of fish and fetishes, or religious beliefs], as the population on them is sparse – the African, like myself, abhorring cool air. Nevertheless, I feel quite sure that no white man has ever looked on the great Peak of Cameroon without a desire arising in his mind to ascend it and know in detail the highest point on the western side of the continent, and indeed one of the highest points in all Africa. […] I have given in to the temptation and am the third Englishman to ascend the Peak and the first to have ascended it from the south-east face. The first man to reach the summit was Sir Richard Burton, accompanied by the great botanist, Gustav Mann. He went up, as did the succeeding twenty-five (mostly Germans) from Babundi; a place on the seashore to the west.

QUESTIONS

Mary Kingsley considered herself a proper scientist and yet her accounts are run through with self-depreciating humour – why do you think she wrote in this way?

The gendering of Kingsley's account is apparently contradictory. On the one hand she extols 'the blessings of a good thick skirt' and discusses the importance of always being properly turned out as an English lady when there was the possibility of meeting other European explorers; on the other, she talks of being the third 'Englishman' to reach the top of a particular mountain. Why might this be the case?

Some feminist geographers have celebrated female travel writers for their subversion of the dominant forms of Orientalist representation. Some have even seen the likes of Mary Kingsley as proto-postcolonialists. These lady travellers were not all commanding, they did not write in a distanced or scientific way, and they were not part of official expeditions sponsored by scientific societies or governments, so they tended to speak to ordinary members of the countries they visited rather than officials. Mary Kingsley argued that it was the old ladies in a village who had all of the knowledge of what was going on and not the village big men. So, did 'lady travellers' produce a better knowledge of the places they had visited, one that avoided the Orientalism contained in the accounts of their male counterparts?

We need to be wary of accepting uncritically the accounts of female travel writers as subversive of Orientalism. One of the great attractions of far-away places to European women travellers was the freedom such places offered. They were able to travel and explore and to have a power and independence that were impossible at home. But they had such power and independence because of their race. They were not just women travellers, they were *white* women travellers. While patriarchy repressed them at home, racism facilitated these women's freedom in the Orient. In addition, because women could not travel as part of government, military or academic societies, they had to find their own funds to finance their travels. As such travel was very expensive, it tended to be wealthy women who were able to explore. Thus, their accounts were structured not only by their whiteness but also by their higher class position – marked not only by Orientalist assumptions about racial difference, but, like many a male traveller, marked also by assumptions related to their wealth and the kind of privileged life they had been accustomed to lead.

When thinking about alternative accounts, it is also important to remember that women were not the only travellers who presented potentially challenging accounts. In any society there are always different forms of masculinity. As well as dominant masculinity (in this case a strong, independent, heroic figure) there are always other expressions of masculinity, although these are often repressed or hidden for fear of reprisals. It is possible then that some travellers challenged the dominant forms of masculinity, more extensively. It is also likely that some were homosexual or bisexual and, through travel, had the freedom to express themselves without the fear of being labelled as an outsider as they would have been liable to at home. This remains quite contested as it was not usually written about directly. In his work on the relationship between geography, exploration and empire, Felix Driver is keen to emphasise the complex motivations and meanings caught up in travel.

Far from being a homogeneous field, then, the culture of exploration was riven with differences over the style, methods and function of the explorer: the quest for geographical truth took many forms. This was partly a matter of audience, for what might be acceptable in one context (a tale of adventure, for example) was not necessarily credible in another (a scientific meaning). It is worth noting here that many of the explorers of the nineteenth century wrote fiction as well as exploration narratives: frequently, they intended the stories for women and children and the narratives for a masculine public. As this suggests, the culture of exploration was profoundly gendered: the man of science was no mere figure of speech. Yet it would be wrong to overemphasize the solidity of the boundary between adventure fiction and exploration narratives: with Stanley, for example, these genres were so blurred that it was difficult to say where one ended and the other began. More generally, the idea of exploration, of travel across the blank spaces, provided a common vehicle for very different forms of practice and different kinds of knowledge. (Driver 2001: 10)

THE WORLD AS EXHIBITION

It was not only through travel writing and travellers' tales that the majority of the European population learnt about empire. Various popular cultural forms drew on images from around the world, whether for education or entertainment. For instance, many advertisers utilised images of empire to sell their products (see Figure 2.3).

World's Fairs and national exhibitions also provided important opportunities for Europeans to view the world that had been conquered by empire builders. These were not simply trade exhibitions but were also manifestations of nations flexing their economic, military, and cultural muscles. The World's Fairs ran from 1851 (the Great Exhibition in London) through to the mid-twentieth century. This was also a time that saw the rise of a museum culture with similar aims, although perhaps less of a sense of spectacle and entertainment.

These World's Fairs celebrated industry in particular and the machine especially (remember the importance of science and machinery to definitions of progress and civilisation). Their principle object was to demonstrate the progress of different countries in one or several branches of production. This was seen as a peaceful way to foster national competition and mutual benefit in terms of the development of global trade. But it also provided a taxonomy of the world, what Timothy Mitchell (1988) has called 'world as exhibition'. This exhibition was of different cultures, different products, different peoples and different colonial possessions. The spectacle of difference offered a convergence of the scientific and the popular, the educative and the entertaining, and gave 'a valuable opportunity for amusement and self-improvement'.

The popularity of these World's Fairs was down to this combination. They were not just exhibitions of raw materials and technology, but also included various dramatic interludes and re-enactments which were literally performances of empire for the home population. Detailed streetscapes and landscapes of

Figure 2.3 Advertising images from the nineteenth century (think about what discourses are drawn upon in these images, what kinds of global geographical connections were being made, and what are the relationships between the product, race and morality?)

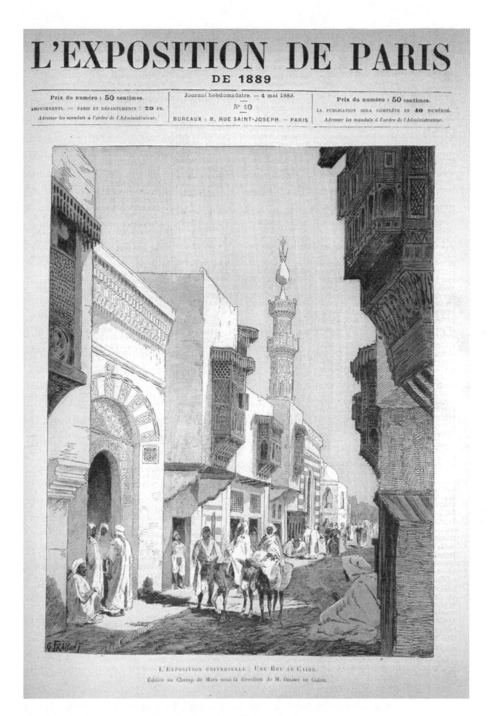

Figure 2.4   Cairo street scene at the Exposition Universelle, Paris, 1889

Entire landscapes were reconstructed to let visitors experience different places. Compare this with the 'reality effect' of the Orientalist paintings discussed in the previous chapter.

different parts of the world were constructed for people to wander through in order to learn about the newest parts of their nation's empire (see Figure 2.4). This allowed the display of information in a way that people would understand. It was structured to inform and entertain, to improve their taste and morals – a kind of amusement without excess. The Eiffel Tower, built for the Paris *exposition universalle* at the end of the nineteenth century, was not only a symbol of the technical achievements of the age but was also a pleasurable distraction. A variety of entry fees was charged (at different times and on different days) to encourage as wide a range of people as possible. The mix of entertainment and imperial propaganda glorified and domesticated empire, making it meaningful for the home populations and giving them a stake in it. The juxtaposition of achievements in science and technology with evidence of colonial acquisitions, alongside the national flag, must have provided important reasons for national identification. National identity was tied in with imperial possession so that World's Fairs represented the greatness of, say, Britain vis-à-vis other countries, thereby showing the population how good it was to be British.

The geography of the fairs was arranged in one of two ways; either colonies were grouped in relation to trade with a metropolitan country, or they were organised in evolutionary terms. In some cases of the latter there was an overt sense of the progression of the human species, so that the pathway through exhibits at the Buffalo Fair in 1898 took visitors on a trip through human time. In others, the juxtaposition was more implicit but no less subtle, as Greenhalgh (1988: 97) indicates for the Paris exhibition in 1931:

> The Eiffel Tower leaned over the site of the [African and Asian] native villages, casting its shadow over them like a giant triumphal monument. A culture/ nature juxtaposition of terrifying simplicity, the vast, gaunt tower represented the power that had enabled the imperial take-over of the lands the villages stood for.

The 1890 French Exhibition at Earl's Court had Arabs act out various scenes, including races, duelling and the kidnap of distressed white maidens. These performances of stereotypes of empire worked to reinforce Orientalist ideas because of the apparent authenticity of the performance (just as the realism of Orientalist paintings had also done).

But it was 'the village' that was of most interest. From 1889 to 1914 people were brought from distant parts of the empire to these fairs to be seen going about their daily business. They were provided with materials to build their own dwellings and raw materials for meals and to produce clothes. There were up to 200 people per village, although the figure was more usually around 50. Again this was important for nation building as well as education and entertainment, as the public saw what 'belonged' to them in a simulated natural state. It was an attempt to provide a total view of life. This drive for authenticity of the object, collected through scientific endeavour and not plunder, was important to the discourses of the fairs (and museums). For example, for the 1901 exhibition in London a Sudanese village was advertised as providing an 'unequalled opportunity of studying the customs of our dusky fellow subjects in their homes and habits, as they could formally be observed only by the African traveller'. For a Zulu village at another exhibition, explorer Henry Morton Stanley endorsed the veracity of the experience:

> Your 'savages' are real African natives, their dresses and dances, equipments and actions are also very real, and when I heard their songs I almost fancied myself among the Mazamboni near Lake Albert once more. (quoted in Coombes, 1994: 88)

Unlike the solidly built pavilions to industry in Europe, the villages were temporary and obviously part of a performance. They were based on everyday life rather than anything more important and lasting, again drawing out the differences between the timelessness of Oriental life and the dynamism of Europe.

The images that some exhibitions provided were, however, more complex. Religious and missionary societies both promoted and criticised imperial and colonial policy throughout the period. They held their own exhibitions to demonstrate the important work being undertaken by religious figures in the colonies. One important reason for this was to raise money! This proved successful as even the poorest in Britain were reminded that however bad their lives were, others were in a worse state of godlessness. In the 'Orient in London' exhibition in 1908 the working classes were called to feel for the injustices faced by others and forget their own. As well as the usual exhibits, there were talks and debates on 'Religion and fetishism in West Africa', 'The liquor traffic in West Africa' and 'Slavery in West Africa'.

Religious exhibitions still presented a spectacle of Africa, the colonial world-as-exhibition, and included such things as displays of Africans at work. But rather different images were projected – these were not of simple savages, but of skilled artisans who would benefit from the teachings of Europeans. Rather than wallowing in timeless backwardness, these were redeemable '**noble savages**' (those who were seen as 'backward' but who had a nobility in their simple lives, and so were regarded as redeemable if they could be guided by Europeans). This image served to underscore the importance of funding missionary works to help develop industry and morality amongst these people.

FRACTURES IN THE REPRESENTATION

As we have seen, one of the critiques of Said's *Orientalism* was that it saw representations of the Orient as being unitary and coherent, changing little over time. Some critics have suggested that this is because of his focus upon the official representations of statecraft, academics and high culture. As the example of female travellers has already suggested, however, the image of the Orient was not quite so coherent or singular. This chapter will conclude with a consideration of both class and national differences. Gender and sexuality were not the only challenges to the singular representation of Orientalism.

Class

Once into the nineteenth century, the middle classes had a greater and greater influence on the dominant representations perpetuated in education and governance. Their lack of understanding of the working classes amassing in the industrial cities led to a fear not dissimilar to the colonists' fear of unknown natives. There was a particular anxiety regarding uneducated working-class people being vulnerable to political agitators, especially communists. It was thus considered important to educate the working classes in the values of the middle classes, particularly sexual morality, self-control and self-improvement. This 'social imperialism' was established through the introduction of museums and the situating of statues in public spaces of those worthy people to whom the working classes should aspire. There was a belief that exposure to objects of civilisation and high art in museums and to examples of civic statuary would improve the working classes, as if by osmosis through close proximity.

While this culture was undoubtedly influential, and is still in evidence in the city centres of European cities today, this was not all that comprised working-class culture – less permanent, and perhaps undocumented sources of culture were also very important. For instance, the music halls of the nineteenth century Britain were incredibly popular and contained humorous and often irreverent songs and jokes about the ruling classes, as well as colonialism and the empire, mixed together with nationalistic passions. While there were certainly Orientalist representations of those outside Europe then, the colonial officers and rulers fared little better. We might consider a twentieth century equivalent to be a film like *Carry on Up the Khyber*, which represented the colonisers as considerably more irrational (and sexually deviant) than the natives (see Figure 2.5). A large part of the film's narrative is organised around challenges to the strong masculinity usually associated with the British colonial project and this is thoroughly sent up through symbolic emasculation via (a) their weapons, (b) their enjoyment of cross-dressing, and (c) a parade which demonstrates that their famous hardiness (wearing kilts without underwear in the freezing climes of the Khyber Pass) is based on a lie. In one of the film's most celebrated scenes, the famed British 'stiff upper lip' is also sent up (d) as a dinner party continues, following all the 'proper' rules of manners and politeness, despite the fact that the building is being shelled.

Figure 2.5 Stills from *Carry on up the Khyber*

Nation

As we have already noted, Said has been accused of Occidentalism in his view of European knowledge. Clearly there are power relations within the governance of Europe, with minor national groups insisting that they are discriminated against formally. Some interpretations of Scotland's and Ireland's past – and sometimes their present – have used the term 'internal colonialism' to discuss their relations with England (and the term is also used by the Basques in Spain).

The nineteenth century was a period of intense nation-building where various states were attempting to incorporate often diverse and historically contrary groups into one modern nation-state. Just as there was a projection of normative culture in terms of class then, this also existed in terms of national identity. The World's Fairs were clearly important for this project as they prominently displayed the successes and achievements of each nation-state alongside others. Their organisers actively encouraged all sorts of groups of people to attend the fairs in order to view these achievements and thus the benefits of being from a particular country.

The Celtic fringe of Britain, for instance, was, to some extent, dealt with in similar ways to the Orient that lay beyond the boundaries of Europe. Just as English anthropologists studied the skeletons and bodily dimensions of the peoples of the colonies, in the same way they catalogued the 'typical' Scottish, Welsh and Irish, in order to demonstrate the supremacy of the English race.

This was also evident in the World's Fairs. The Celtic fringe was represented in ways that were not dissimilar to the ways in which the peoples of the British Empire were shown. Ireland was presented as ancient and rural, with thatched cottages, traditional dancing and use of the Gaelic language, a romantic image of happy self-sufficiency not so different from the African villages assembled nearby. Similarly, despite the booming industry of Glasgow, 'second city of the empire', Scotland was represented as a highland idyll. In the context of nation-building, this played an important role, as potential political differences between the nations making up Britain were rendered no more than aesthetic and folksy.

Further reading

On colonial knowledge

Adas, M. (1989) *Machines as the Measures of Man*. Ithaca, NY: Cornell University Press.
Arnold, D. (1993) *Colonizing the Body*. Berkeley, CA: University of California Press.

On travel writers

Blunt, A. (1994) *Travel, Gender and Imperialism: Mary Kingsley and West Africa*. New York: The Guilford Press.
Driver, F. (2001) *Geography Militant: Cultures of Exploration and Empire*. Oxford: Blackwell.
Pratt, M.L. (1992) *Imperial Eyes: Travel Writing and Transculturation*. London: Routledge.
Riffenburgh, B. (1993) *The Myth of the Explorer*. London: Wiley.

On the role of women travellers in geography

Domosh, M. (1991) 'Towards a feminist historiography of geography', *Transactions of the Institute of British Geographers*, 16: 95–104.
Stoddart, D. (1991) 'Do we need a feminist historiography of geography – and if we do, what should it be like?', *Transactions of the Institute of British Geographers*, 16: 484–87.

On the popularisation of empire

Greenhalgh, P. (1988) *Ephemeral Vistas: The Expositions Universelles, Great Exhibitions and World's Fairs, 1851–1939*. Manchester: Manchester University Press.
McClintock, A. (1995) *Imperial Leather*. London: Routledge.
MacKenzie, J. (ed.) (1986) *Imperialism and Popular Culture*. Manchester: Manchester University Press.

3

LANDSCAPES OF POWER

The Indian or native city is situated often at a considerable distance from the European civil lines and military cantonments, in one or the other of which the Europeans live. [It] is usually walled. The houses are closely packed together, the streets being very narrow ... Even the main street, in which the chief business is transacted, will hardly allow of one cart passing another. The houses are high and most picturesque, though very dirty. The bazaar is a feast of colour. The booth-like open shops filled with many-hued wares, gay silks and cottons, and piles of luscious fruits, with the brilliantly coloured garments of the passers-by and of the loungers (for in the East there is no hurry), make the native city a joy for the lover of colour. The effect would be garish, but with the background of closely set fantastic buildings, the sunny lights and deep velvety shadows, the picture gives joy and satisfaction to the onlooker. True, a captious critic does not approve of what he sees on close inspection, and the state of sanitation is such that diseases when introduced spread with incredible rapidity. It is not without reason that the European residential quarter is built at a considerable distance from the fascinating but dangerous native city. (Platt, 1923, quoted in King, 1976: 127)

The form of the colonial built environment can tell us a good deal about how the colonisers viewed the native people and their landscape, but we can also see what happened when the ideas of the colonisers were put into practice – the outcomes were not always what they expected! As we can gather from the opening quote, native towns were often viewed with a mixture of wonder and fear. The colours and designs were seen as exotic and exciting, but at the same time, their apparent lack of order and design was worrying to the colonisers. As we know from the previous chapter, colonial systems were powerful as a result of the knowledge created about native peoples. The apparent illegibility of the native quarters was therefore a threat to Europeans. Sometimes native areas were ordered and rebuilt; most often, though, European quarters were established in distinct areas, sometimes at quite a distance from the natives.

POWER IN THE LANDSCAPE

When it came to the landscape, colonialism was about transformation. Just as colonial knowledge sought to order the world in a taxonomy of the known, the engineers of the colonial landscape sought to order the colonies into a knowable pattern. Colonial landscapes were ordered, sanitised, made amenable to regulation, and structured to enhance the flow of economic activities. Thus, these landscapes did not simply reflect colonial aspirations but were also both consciously and unconsciously used as social technologies, as strategies of power to incorporate, categorise, discipline, control and reform the inhabitants of the city, town or plantation. It was therefore intended that the use of buildings and the urban form itself would start affecting the nature of native populations.

Components of such a framework of colonial power in the landscape varied between the cities of the different colonial powers in place, but included:

- The church (especially in Latin America).
- Trading companies such as the British East India Company.
- The military (which wanted an ordered, visible landscape that was easy to control).
- The colonial state itself.

Through the actions of these institutions, colonial policy was made concrete through colonial space and practice.

In many cases, the first stage in the colonial process was the seizure of land belonging to the previous ruler and using this to weaken the power of native institutions. This simultaneously undermined the economic base and attacked symbolic power. There are many examples throughout the world of conquerors taking over symbolically important sites, whether these were religious or political. For instance, the Spanish *conquistadors* established their capital, Mexico City, on the ruins of the Aztec capital Tenoctitlan. They destroyed the indigenous city and established a new one in its place as a marker of power – their power over the old order. At other times, colonists have propped up the elements of traditional power once they have been conquered in order to help establish firm colonial rule. In Morocco, French colonists rebuilt the palace of the Sultan in Rabat but the governor's residency was placed next to it, drawing on the symbolic power of this association.

THE POWER OF THE LANDSCAPE

The landscape then is one further form through which discourses about European mastery could be expressed. James Duncan argues that the landscape is particularly important because it works to help make certain values seem natural (self-evidently true rather than someone's opinion) because it seems non-political. The landscape is just there. However, encoded within the landscape are particular values. Consider the following quotes from Duncan:

> The landscape is a text in the language of built form which is explicitly read or subconsciously apprehended by those who live and work within its presence. The power of landscape features lies in the fact that they are easy to grasp both emotionally and intellectually, for they can be visited, touched, venerated, and often most importantly, taken for granted as right and natural. (1992: 81)

> The landscape, I would argue, is one of the central elements in a cultural system, for as an ordered assemblage of objects, a text, it acts as a signifying system through which the social system is communicated, reproduced, experienced, and explored. (1990: 17)

> ... the landscape of the city was a political tract written in space and carved in stone. The landscape was part of the practice of power. (1992: 86)

How did power work through the colonial urban landscape?

First we need to turn to the work of Michel Foucault, and in particular his conceptualisation of the **panopticon**. This was based upon an intriguing prison design by Jeremy Bentham (Figure 3.1, p. 58). The prison was designed around a central tower. The prisoners were housed in individual cells around the outside walls of the prison. While these were visible from the centre of the prison, there were walls between individual prisoners that made contact between inmates difficult. The guards' tower was in the middle. This had tiny slit-like windows which allowed them to view the prisoners, but made it impossible for the prisoners to see when they were being watched by the guard. This meant that they had to behave at all times *in case the guard was watching*. But what intrigued Foucault most was the fact that as there was no way of the prisoners knowing when they were being watched, there was not in fact any need for a guard to be there at all. It was the structure of the building that facilitated the disciplining of the prisoners, and the gaze of power which emanated from that central tower that ensured self-discipline. Foucault took this as a metaphor for the way that modern societies worked. Rather than having to resort to violent means of control as earlier societies had (hangings and other gruesome public punishments for those who disobeyed the ruler), it gave the possibility of constant surveillance throughout people's lives, from school through to work, and the establishment of institutions of knowledge collection run by the state.

Thus, in terms of the colonial landscape we have the operations of power working subtly through the landscape. For instance, the layout of colonial plantations was such that the workers' accommodation and the rows of crops would sometimes radiate outwards from the owner or manager's residence,

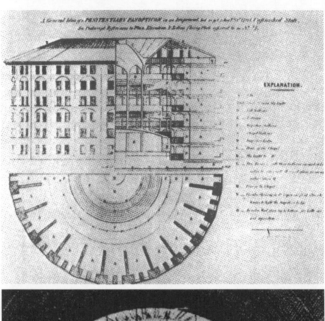

Figure 3.1 Bentham's panopticon

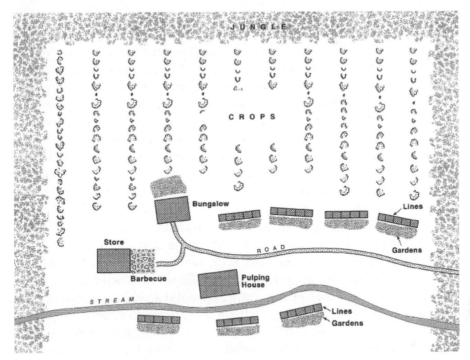

Figure 3.2 Schematic diagram of a coffee plantation, mid-nineteenth century. The 'lines' are workers' dwellings

or would be ordered along easily monitored straight lines (see Figure 3.2). This allowed surveillance of the workers from a central point, a bit like the guard in the panopticon tower, and once again it would not always have been apparent to the workers whether they were being watched or not.

Foucault's metaphor for modern society has been highly influential but post-colonial scholars have suggested that it is a Eurocentric model and has to be seen in a more partial way. For while it is convincing to argue that panopticism has become dominant in western societies, under colonialism the ruling powers still continued to use spectacular forms of power (such as publicly violent punishments) alongside the subtleties of a disciplinary landscape to maintain control.

Therefore, we also need to consider other understandings of how power worked through the colonial landscape. Georges Bataille, another French philosopher of the built form, turns Foucault on his head. Rather than power running silently through all relations, for Bataille power can be visualised in the landscape. Applied through colonial architecture, locals are meant to see power openly reflected in the monumental architecture of the colonisers. Architecture acts as a metaphor of the power of the colonisers and it is intended to overawe its new subjects. The gaze here is the reverse to Foucault's model, as it is the subjects who gaze at the signs of power in the architecture, rather than the built form facilitating a disciplinary gaze on the subjects.

Most often civic buildings in colonies were built to impress the colonised, and to imprint firmly the power of the new rulers into the landscape. However, it should be noted that such architecture tended to be located within the capitals and major cities and that architecture elsewhere tended to be much more modest and mundane. Colonial architectural investments also differed by location. The British invested more time and money to changing the Indian landscape, for instance, than they did in most of their African colonies, reflecting the relative values they held of each place.

In India the British created landscapes of power in a new capital. New Delhi was planned to incorporate spectacular and grand architecture to enforce British dominance. This plan included a conscious attempt to civilise the Indian population. The architect of this plan and the major buildings, Edward Lutyens, had nothing but contempt for Indian architectural heritage. He considered that India had no architecture before the arrival of Europeans, just tents in stone (Irvine, 1981). He, and many other European architects of the time, interpreted Islamic art as feminine, insufficiently structured, and exotic and imaginary rather than practical. He assumed that architectural styles had remained constant for all time rather than progressing and developing as had been the case for European architecture. British architecture was supposed to 'improve' the natives, especially via public buildings and museums. The order and structure of the streets, the good examples made by those celebrated in public statuary, and the informative and enlightening culture of museums, were all seen as having a positive effect on colonial people – the cultured nature of such built forms was expected to 'rub off' on the morality and values of colonial viewers. As we have already seen, the Victorian bourgeoisie had the same plans back at home for the working classes, whom they considered were in need of the same moral guidance and enlightenment.

Clearly it was an extravagant move to build a new city from scratch but the British colonisers were facing nationalist stirrings. The Indian Congress Party had been formed in 1885 and had developed an increasingly vocal opposition to colonisation that had lasted into the twentieth century. In response, the British began New Delhi in earnest in 1913. The city was structured around two triumphal avenues, Kingsway and Queensway. The plan offered a clear sense of order, with the main street running between the governor's residence (occupying an elevated position overlooking the general population) and various government buildings. Statues of lions at the entrance to the Viceroy's house were further symbols of Britishness and strength. War memorials were included in this landscape but none of these marked battles between the Indians and the British. Instead, memorials were built to Indian deaths in wars that Britain and her empire had fought with others. This was clearly an attempt to enforce a common identity in the empire; not quite a brotherhood, but a family with Britain as both mother and father, and the colonised people as children under Britain's guidance.

More generally, a colonial town model emerged which involved two separate settlements comprising each city – the Oriental and the western quarters.

Figure 3.3 Contrasting Delhi streetscapes. Top – dense streets of the older part of Delhi. Bottom – the 'ordered' landscape of New Delhi.

These were kept separate by colonial administrators, nominally for health and safety reasons, and protected Europeans from what were believed to be disease-ridden traditional quarters, and also the perceived threat of violence. However, these two parts had to be close enough to allow workers to travel to service the Europeans and also to allow for military control. The Oriental quarters were perceived to be maze-like by the European colonisers who felt that their own urban design reflected order and rationality. There were often very direct interventions in the native quarters, which demonstrated the importance of order as a practical as well as a moral concept. Kumar (2002: 95) explains what happened in the instance of nineteenth century colonial Madras:

> The Inspector of Nuisance (an Indian) was appointed to determine levels of hygiene on a daily basis, as well as to educate the natives in cleaning, personal hygiene and in the prevention of nuisance. It was believed that: 'lessons they would teach of the advantages of obedience to a few simple sanitary laws would in course of time lead the people to adapt of themselves measures calculated to place their village communities under improved sanitary condition' (Ranking, 1869: 2). The Collector noted that: 'The Inspectors I would appoint to certain groups or circles of villages … This class of men need not be highly educated. They would merely be required to know what are 'nuisances' to spy them out and report them' (ibid.:4).

> […]

> Nuisance Inspectors were under the paternalistic purview of the Superintendent of Lands (a European) who was explicitly referred to as a 'Moral Agent' to check evil misuse of power by native subordinates, illustrating the conflation of the moral and the sanitary types of order by the colonial authorities.

This ordering of the landscape had two main purposes. On the one hand, in a very practical sense, it was easier to police and defend, but on the other hand, this had more philosophical grounds: it reflected the rational scientific order that Europeans saw as characterising western thought, as we saw in the last chapter.

As geographers, it is important to bear in mind the landscapes upon which the colonisers were working. Although these were based around colonial plans of how the landscape should operate, such plans did not always work out perfectly in real life. While imperial architectural styles were manifestations of interconnected structures of power/knowledge that informed colonialism everywhere, there were always certain constraints and limitations placed by the materials and wealth available and by the environmental conditions. Much of the postcolonial critique of colonial practice has emerged from literary and cultural studies which have tended to focus on plans and documents rather than the everyday practices of colonialism (something which we will discuss towards the end of the book). However, outside of the capital cities, much colonial life was mundane and unspectacular with architecture that had had to adapt to local conditions and available resources. Just because the colonisers were trying to impose an image of order, this was not always completely successful. In her work on colonial Singapore, Yeoh (1996: 10, 13, 15) explains:

> The colonial urban landscape is hence not simply a palimpsest reflecting the impress of asymmetrical power relations undergirding colonial society, but

also a terrain of discipline and resistance, a resource drawn upon by different groups and the contended object of everyday discourse in conflicts and negotiations involving *both* colonists and colonized groups. It embodies the negotiation of power between the dominant and the subordinated in society, each with their own versions of reality and practice.

[...]

To attribute an absolute *omnipotence* to the 'apparatus' of disciplinary power would, as Colin Gordon has argued, confuse the domain of *discourse* with those of *practices* and *effects* for what is intended and articulated by the 'powerful' within the domain of discourse (such as those of sanitary sciences or urban planning) may fail to materialize in its entirety when transposed to the domain of actual practices and techniques or produce unintended consequences and effects. It would also amount to denying that those which disciplinary power seeks to control are capable of counter-strategies which can challenge disciplinary power and modify its effects.

[...]

These counter-strategies included 'active' forms such as rioting, holding demonstrations, or going on strikes as means of expressing grievances. However, more common than 'active' and 'heroic' forms of protest, and in the long run less costly in terms of effort and sacrifice, were the 'passive', rather unspectacular means of countering and inflecting colonial control. The community could adopt an outward attitude of apparent acquiescence (or at least non-protest), but in reality disregard or even thwart the measures imposed by the colonial power. They could drag their feet over or dig their heels into requests for co-operation. [...] Compliance was withdrawn unobtrusively, without calling attention to the act itself or upsetting the larger symbolic order of dominance and dependence prescribed for the colonial world. Even when such forms of resistance became widespread enough to awaken the colonialists to the inefficacy of their policies at grassroots level, they were often too dispersed, too anonymous, and far too commonplace to allow immediate effective action against the actual culprits.

The colonial urban landscape was hence not simply a surface reflecting the effects of the unequal power relations characterising colonial societies, but also a resource drawn upon in conflicts involving *both* colonists and colonised groups. It embodied the negotiation of the power of domination and resistance between the dominant and subordinated in society, each with their own versions of reality and practice. While the colonised may have behaved themselves in public and acquiesced to rules and regulations, this does not mean that they internalised the meaning encoded in the landscape. In the safety of private spaces, in their homes or with friends, they might have discussed rather different interpretations, challenging or poking fun at the images imposed by the colonisers.

It is important to note here that the notion of power running through the landscape in the ways expressed above is not something that is relegated only to the distant colonial past. The use of the landscape to emphasise the power of one group over another was perhaps at its most blatant in South Africa during apartheid. From 1948 until its repeal in the 1990s, apartheid operated following three spatial scales – the personal, urban residential and national – and was designed to keep the races (white, black, coloured and Asian) apart, in terms of differential access to public facilities or residences. Apartheid thus also promoted differential access to resources and life-opportunities.

ABSTRACT SPACES: THE COLONIAL LANDSCAPE, WORK AND THE BODY

The creation of maps and plans of colonial territory, and the establishment of ownership over this land, had the effect of creating what Henri Lefebvre (1991) has called '**abstract space**'. Lefebvre observed that to control the production of space is to control the processes of social production and reproduction. Central to this is the commodification and bureaucratisation of everyday life, namely making space mathematical and ordered (challenging the indigenous ordering of space) in such a way as to render the colony most efficiently known and governable. As we have already seen, geographers were among those who did just this; producing maps of colonised territories, measuring distances, noting landownership, and so on. This was a process that fed into the colony as a site of production: the selling and taxing of land (where perhaps the notion of land ownership had previously not existed) as well as assisting the colonisers' rule through the knowledge this produced about the area.

The links between capitalism and modernity have been explicated by many authors. But as capitalism spread to colonies, so modernity became part of the imperialist project and cultural transformation became based on imperialist knowledge. Duncan (2002) has argued that abstract space requires the construction of 'abstract bodies' to conform to it. Abstract bodies are bodies that are docile, useful, disciplined, rationalised, normalised, and controlled sexually. In short, they are economic investments to be protected and utilised to their greatest capacity.

In his work on colonial plantations, Duncan (2002) develops the ideas of the production of colonial urban space into non-urban elements of the colonies. He argued that the production of abstract bodies in a place like Ceylon or other colonies required the cultural transformation of a people, by attempting to create a new de-cultured worker who could labour productively in the colonial plantations.

Plantation owners' ideas were based on a nineteenth century belief in scientific solutions to what were seen as highly interrelated problems of race, moral depravity, disease, material squalour and political disorder. They tried to transform what was seen as the flawed native body into the abstract body of the labourer, a body that corresponded to abstract routines of labour in time and space. Here is what the Chief Medical Officer in Ceylon in the 1870s had to say about local workers or 'coolies': 'The coolie is naturally lazy, indolent and docile ... He has strongly developed animal passions' (quoted in Duncan, 2002: 324). In a despatch to the Colonial Office, Governor Robinson claimed that

> The Tamil coolie is perhaps the simplest, as he is certainly the most capricious, of all the Orientals with whom we have had to deal in Ceylon. He is like a child requiring the strong arm of power. He must know that he is subject to paternal authority. (quoted in Duncan, 2002: 324)

Through the ideas of control and rationality covered in the previous chapter, plantation owners believed that work routines could be established to create effective labourers. The need to turn the 'indolent, pleasure-addicted, sensuous peoples' of the Orient into efficient labourers 'inspired innumerable discourses on the techniques of supervision and control' (Adas, 1989: 258). Work routines were centred around the notion of a flawed native body, compared to the abstract, ideal, European body. This view emerged from the belief that native peoples were closer to nature than Europeans that we have come across already. So, for example, wages were determined by bodily characteristics: in the 1870s, men were paid 9d, women 7d, and children by their height. Plantation owners argued that it was necessary to limit wages for the good of the labourers. Because of their closeness to nature, their animal passions and needs only required subsistence wages. They merely needed to support their own bodily requirements, it was believed, so to give them more money would be reckless. Workers were to be physically disciplined if they did not work. Punishments were corporal because of the embodied nature of the workers. Thus, the native workers were perceived to be improved through their work on the abstract space of the plantation:

> Plantations can be conceived of as modern technologies for the reconfiguration of space, tools, scientific instruments and other material resources, bringing together culturally heterogeneous populations, stripping them of their former social attachments and reconstituting them as workers through the use of space-time strategies of monitoring and control. (Duncan, 2002: 317)

The native body was regarded as an instance of nature which was worked upon through labour and the abstract space of the plantation to create the abstract body of the worker. Thus, labour westernises the native body. If we return to the views of the Chief Medical Officer of Ceylon, we can see this view very clearly:

> It is "genius of labour," he said, which transforms the coolie. His countenance "mirrors the newly awakened soul, and the consciousness of powers and capabilities, hitherto dormant and unused, stamps the physiognomy with an expression of manliness and intelligence which is never seen in the raw, uncivilised, newly landed coolie". (quoted in Duncan, 2002: 326)

Imperialist knowledge was command over time and space – it was rationality, order and self-control – all things which natives lacked and colonials possessed. By enforcing a work discipline onto the natives (providing an ordering of time and space in the abstract space of the plantation), the plantation would transform the native from a culturally-marked body into a labourer, and capitalism could be spread throughout the colonies. Of course, this was very convenient because it suggested that hard work was beneficial to the colonised people (rather than being seen as a burden to them, as is a more usual interpretation). Therefore, this discourse presented the needs of the colonials and the good of the natives to be the same.

LANDSCAPES OF HOME

Part of power is to be able domesticate the unfamiliar, in other words, to create home in distant and foreign places. When building cities and residences, colonists drew upon the styles with which they were familiar, which they understood and which they valued. When in the colonies, Europeans reproduced home life but with a difference. Race stood in for class. Colonists could act out a higher class status in colonies so they could reproduce the landscapes of home – but with added luxury. Europeans did not want to go to the colonies to live as they did back home, but wanted to improve their lot. This meant that European-style architecture was established in the colonies, even when at odds with the local style and environment (see Figure 3.4).

Clearly, in many parts of the empire, climate was a great drawback to recreating home. Buildings and lifestyles had to make compromises to deal with searing heat and humidity. In South Asia, this was negotiated through the creation of hill stations, built from 1819 onwards as high elevation retreats from heat, dust and the natives. By the 1850s the colonial government had established alternative seats of government for the hot summer months. In addition to being a more pleasant environment for Europeans, it was thought to save administrators from the degenerative effects of the climate.

There were also more directly related health issues. The colonisers considered all natives to be unhealthy and this in turn warranted segregation, as we saw in terms of the colonial cities. Disease was thought of in moralistic terms as well as in a literal bodily sense. Hill stations were ideal because of the healthier climate and being away from Indian population densities. They were 'comforting little pieces of England' as one contemporary visitor said (quoted in Kenny, 1995: 711).

This was perceived to be especially important for women who, because of their closer proximity to nature, were thought to be more affected by the extremes of the Indian climate. To ensure that women (and their morals) did

Figure 3.4   French style architecture in Hanoi, Vietnam

not fall victim to the heat, they periodically needed to escape to the hill stations. But this environment also made it possible to introduce European plants, European housing, clubs, pastimes, and to produce a more convincing and recognisable landscape of home. Plantations were given English names such as Eton, Pine Hill, Gloucester, and Wiltshire. Plantation owners adopted a model of the country gentleman. In 1877 one commentator said of Ootacamund in the South of India: 'I affirm it to be a paradise … The afternoon was rainy and the road muddy, but such *English* rain, such deliciously

English mud' (quoted in Kenny, 1995: 702). The landscape of the hill stations was so familiar to the English that they were almost accepted to *be* part of England.

Again we can see the power of colonialism to write meaning onto the native landscape. Here the colonialists were able to domesticate the different landscape to render it in a form that was familiar and known to them. It allowed them to feel at home in such foreign lands (and perhaps also made the natives feel out of place in their own lands).

Further reading

On the construction of the colonial landscape

Carter, P. (1987) *The Road to Botany Bay: An Exploration of Landscape and History*. New York: Knopf.

King, A.D. (1976) *Colonial Urban Development*. London: Routledge.

Mitchell, T. (1988) *Colonising Egypt*. Berkeley, CA: University of California.

Myers, G. (2003) *Verandas of Power: Colonialism and Space in Urban Africa*. Syracuse, NY: Syracuse University Press.

Yeoh, B. (1996) *Contesting Space: Power Relations and the Urban Built Environment in Colonial Singapore*. Oxford: Oxford University Press.

On the creation of abstract space and abstract bodies

Duncan, J. (2007) *In the Shadows of the Tropics: Climate, Race and Biopower in Nineteenth Century Ceylon*: Aldershot: Ashgate.

Duncan, J. (2002) 'Embodying colonialism?: Domination and resistance in 19th century Ceylonese coffee plantations', *Journal of Historical Geography*, 28 (3): 317–38.

Lefebvre, H. (1991) *The Production of Space*. Oxford: Blackwell.

On the domestication of colonial landscapes

Blunt, A. (2005) *Domicile and Diaspora: Anglo-Indian Women and the Spatial Politics of Home*. Oxford: Blackwell.

Kenny, J. (1995) 'Climate, race, and imperial authority: the symbolic landscape of the British hill station in India', *Annals, Association of American Geographers*, 85: 694–714.

A film that conveys a sense of the mundane and unspectacular, everyday nature of much of the colonial experience, but also the highly symbolic nature of the landscape, is Jean-Jacque Annaud's *Noirs et Blancs en Couleur* (1976), released in English on DVD as *Black and White in Color* (2003).

PART II

Post-Colonialisms

As already indicated, in conventional accounts, decolonisation marked the end of colonialism. However, postcolonial approaches see continuity from the colonial to the post-colonial periods. While the idea of neo-colonialism (literally a 'new' colonialism) has long been argued for in economic terms – in that there are still relationships of dependency between ex-colonies and the powers that had once ruled them – here we will examine the continuing legacies of colonial ways of knowing. In Chapter 5 we will consider the shape of the post-colonial world order to see if it has indeed moved away from the binary of west-rest that Said and others witnessed during the colonial period. We will also think about American ideas on development as a replacement for colonial relations, and the rise of the Third World. In Chapter 6 we will see how the exotic and the other – concepts at the heart of colonial travel writing and exploration – have been reformulated in the post-colonial era where all parts of the world are known and explored, and will examine the ways in which they have become incorporated as products in the world economy. While this chapter will consider arguments about globalisation and cultural homogenisation, it will also argue for the ways in which different parts of the world are resisting this process and offering alternative expressions of cultural belonging and identity.

4

NEW ORDERS?

Political sovereignty is but a mockery without the means of meeting poverty and illiteracy and disease. Self-determination is but a slogan if the future holds no hope.

That is why my nation, which has freely shared its capital and its technology to help others help themselves, now proposes officially dedicating this decade of the 1960s as the United Nations Decade of Development. Under the framework of that Resolution, the United Nations' existing efforts in promoting economic growth can be expanded and coordinated. Regional surveys and training institutes can now pool the talents of many. New research, technical assistance and pilot projects can unlock the wealth of less developed lands and untapped waters. And development can become a cooperative and not a competitive enterprise – to enable all nations, however diverse in their systems and beliefs, to become in fact as well as in law free and equal nations.

My country favors a world of free and equal states. We agree with those who say that colonialism is a key issue in this Assembly. But let the full facts of that issue be discussed in full.

On the one hand is the fact that, since the close of World War II, a worldwide declaration of independence has transformed nearly 1 billion people and 9 million square miles into 42 free and independent states. Less than 2% of the world's population now lives in 'dependant' territories.

I do not ignore the remaining problems of traditional colonialism which still confront this body. These problems can be solved with patience, goodwill, and determination. Within the limits of our responsibility in such matters, my country intends to be a participant and not merely an observer, in the peaceful, expeditious movement of nations from the status of colonies to the partnership of equals. That continuing tide of self determination, which runs so strongly, has our sympathy and our support.

But colonialism in its harshest forms is not only the exploitation of new nations by old, of dark skins by light, or the subjugation of the poor by the rich. My nation was once a colony, and we know what colonialism means; the exploitation and subjugation of the weak by the powerful, of the many by the few, of the governed who have given no consent to be governed, whatever their continent, their class, their color.

President John F. Kennedy, Speech to the United Nations,
25 September 1961, New York

This chapter will consider the cultural ways in which the end of the colonial period was understood. It will focus particularly on fears about the 'closing' of the world, which takes us back to Joseph Conrad's thinking about the romance of difference, and to the rise of new understandings about the relationship between the west and the rest, especially the concept of 'development'. This period saw the shifting of dominance from Europe to the USA and was heralded by an optimism about the possibilities for a better world. However, how much did really change?

THE END OF BLANK SPACES ON THE MAP

In the early twentieth century, fears arose about the fact that all of the world had been colonised and had become known to Europeans because this meant that there were no more dark spaces on the map to explore, conquer or convert. After the 'scramble for Africa', where Europeans raced to lay claims on the continent from around 1880 until the First World War, there was little territory for Europeans to compete over. As we have seen at various points throughout the book so far, colonialism was driven by issues of reason and science, but throughout this it was also shot through by issues of romance, whether for the 'noble savage' or the mysteries of unexplored places. Romance is designed to satisfy the desire for adventure, danger, excitement and otherness. As Said has explained so convincingly, the Orient provided this space of romance for Europe for centuries but there was a feeling that by end of the nineteenth century, with the global reach of European colonialism achieved, that this had come to an end. There were no blank spaces left on the map, no unknown dangerous places within which heroes could prove themselves, no unknown others to meet.

With exploration and then the apparatus of colonial rule the world was divided into the safe, known world and that of the non-west, with the latter characterised by magic and mystery, disorder and otherness yet to be discovered. Processes of colonialism and imperialism open up otherness by domesticating and ordering it. This, however, rids it of its otherness, exoticism and excitement. By end of the nineteenth century it seemed that this romance had come to a close and we saw the emergence of a theme that will run through post-colonial culture: a sense of loss. Global modernisation was being driven by colonialism and was eradicating otherness and exotic 'elsewheres'. Some then saw imperialism as the threat to the romance of otherness, bringing homogeneity and safety to all places. They perceived that the adventure had gone.

For example, if we return to Joseph Conrad looking at maps of Africa, we can see that he epitomised this disappointment by arguing that as everything was named and known – all the blank spaces on the map had been filled in – this had robbed the continent of its mystery:

> Now when I was a little chap I had a passion for maps. I would look for hours at South America, or Africa, or Australia, and lose myself in all the glories of exploration. At that time there were many blank spaces on the earth, and when I saw one that looked particularly inviting on a map (but they all look like that) I would put my finger on it and say: When I grow up I will go there. The North Pole was one of these places, I remember. Well, I haven't been there yet, and shall not try now. The glamour's off. Other places were scattered about the Equator and in every sort of latitude all over the two hemispheres. I have been in some of them and … well, we won't talk about that, But there was one yet – the biggest – the most blank, so to speak – that I had a hankering after.

> True, by this time it was not a blank space any more. It had got filled since my boyhood with rivers and lakes and names. It had ceased to be a blank space of delightful mystery – a white patch for a boy to dream gloriously over. It had become a place of darkness. (Joseph Conrad, *Heart of Darkness*, 1899)

Whether for better or worse (the opening optimism of John F. Kennedy or the disappointment of Conrad), the end of the colonial period seemed to herald a change in the world order. On the surface it would appear that the binary structure had gone and instead post-colonialism saw a shift into a new world order where the Orient was refigured as the Third World. And yet, as we have seen, Said argues that the binary geography of Occident-Orient has persisted to the present day. In this chapter, we will first consider the conventional stories about decolonisation and then seek to understand the nature of the emerging post-colonial world order.

DECOLONISATION

The colonial state relied heavily on the power of knowledge as discussed in Chapter 3 – the idea of knowing the natives in order to control them. This was achieved through the establishment of a modern state which created order within the space of the colony: through defining and classifying the space in maps and patterns of land ownership, by counting populations in censuses, by standardising weights and measures to make trade and commerce efficient and rational and through learning about local languages and customs. This micro control of the state took a great deal of daily work;

dreary middle-management jobs that did not attract the Europeans who travelled to the colonies. Thus, in certain contexts – primarily south Asia – schools were run for the native elite in order to train them to run the lower and middle level governance of their country. This reinforced and naturalised the colonisers' values as right and proper.

Ironically, it has been argued that this process facilitated the establishment of anti-colonial movements. The establishment of colonial governance required the introduction of civil services, print capitalism, a unified language of state and education, railways and maps. While this was vital for effective and efficient governance by the colonial power, it also facilitated the imagination of an alternative – a post-colonial **national identity**. For the first time, it allowed a previously diverse or disparate population, or at least the elite members of that population, the possibility of imagining themselves as a community with a common interest, history and identity. This was particularly the case in India where British colonisers linked the country together physically through the introduction of the railways and linguistically through the introduction of English as a common language of governance. Maps of India allowed people in previously unconnected places to see their common predicament; the English language allowed the various elites to communicate more easily where they had previously been hindered by the existence of many different indigenous languages; the railways allowed news and experiences to travel across the continent. Education taught the Indian elites European concepts of reason. They were educated to believe in Enlightenment values, values that they wanted applied to themselves and their own people. The colonisers had introduced a conceptual and actual language through which these elites could articulate their case for independence.

This is the conventional story. Are we happy with this story?

There is certainly evidence, particularly for India, that this did indeed happen. However, there are those who have challenged this version of events, including Indian historian Partha Chatterjee. Chatterjee has argued that the problem with this explanation of events is that it is still the west that is the hero of the story. It was the enlightened values of the western colonisers, communicated to the indigenous elite, which stimulated the end of colonial rule. The colonised society remains passive in this story, not only awaiting the arrival of the colonists to 'develop' their society and bring reason in the first place, but also to provide the mechanisms of resistance to this regime to hasten its end. Chatterjee (1993: 5) suggests that it is as if the Europeans

> … have thought out on our behalf not only the script of colonial enlightenment and exploitation, but also that of our anticolonial resistance and postcolonial misery … Even our imaginations must remain forever colonised.

Instead, Chatterjee offers a different version of the story wherein the colonised society creates its own domain of knowledge, value and sovereignty before establishing a challenge to colonial power. He suggests that under colonial regimes, social life was divided into two domains:

1 *The material.* This was the public space outside the home where the economy, statecraft, and science and technology were dominated by the colonial powers. Colonised leaders recognised the superiority of the colonisers in this sphere and studied this knowledge with the intention of replicating it.
2 *The spiritual.* This was the private space of the home and the traditional cultures of the colonised society which were repressed during colonialism. Such values have been preserved in the hidden spaces of private life (out of sight of the colonial powers), in the face of colonial attempts at change and, argues Chatterjee, have also been nurtured as the basis for anticolonial resistance.

It is in the spiritual domain that 'nationalism launches its most powerful, creative, and historically significant project: to fashion a "modern" national culture that is nevertheless not Western' (Chatterjee, 1993: 6). This leads to hybrid forms by mixing European ideas of governance and democracy with the cultural values of the spiritual domain ('**hybridity**' is the mixing of two separate things to make a new, third thing, and is a very important concept in postcolonial theory and culture, as we shall see later in the book). However, this is not an argument favoured by all postcolonial theorists. Frantz Fanon warned against the 'pitfalls of national consciousness' wherein the indigenous elites, having absorbed too much of colonial ways, simply replaced the colonisers at the top of the social, political and economic hierarchy. This ensured that for the vast majority of the population, there were no significant changes to the real conditions of existence. This is something we see replicated throughout sub-Saharan Africa and elsewhere. (We will discuss Fanon's more radical arguments later in the book.) Various political alternatives have been considered and attempted as a way of organising postcolonial societies (see the box below).

But the global geography that has come to dominate imaginations is one apparently different from the binary geography of the colonial period – the idea of the three worlds. But as we will see, the logic is not so different from the Orientalism that had dominated the colonial period.

THE THIRD WAY AND TRICONTINENTALISM: ALTERNATIVE WORLD ORDERS

Initial attempts at an organised, alternative, post-colonial world order were first seen in the 'non-aligned movement' of the 1955 Bandung Conference at which there were representatives from a number of Asian and African countries, most notably India, Egypt,

(Cont'd)

Indonesia and Ghana. They sought to establish an alternative political, economic and cultural organisation rooted in the experience of Africa and Asia, presenting a 'third way' which was modern but which followed neither the capitalistic US model, nor Soviet socialism. Many would see the Bandung Conference as marking the beginning of postcolonialism as a conceptual position and as a form of politics.

In Havana, in 1966, a more radical version of postcolonial political philosophy was articulated at the Tricontinental Conference. Robert Young (2003) argued that tricontinentalism is a more appropriate term than postcolonialism, as it describes something that it *is* (the bringing together of the interests of the three continents of the global south; Asia, Africa and Latin America) rather than what it *is not* (after colonialism). Delegates attended from Guinea, the Congo, South Africa, Angola, Vietnam, Syria, North Korea, the Palestinian Liberation Organisation, Puerto Rico, Chile and the Dominican Republic. From this conference emerged a journal, *Tricontinental*, published by the Organisation of Solidarity with the People of Asia, Africa and Latin America (OSPAAAL), and which brought together both theorists and activists including Frantz Fanon, Che Guevara, Ho Chi Minh and Jean-Paul Sartre.

Other post-colonial political organisations and identities operated at the continental scale. Pan-Africanism sought self-determination for African people, arguing that it was meaningless for one African country to achieve its independence while other countries were still colonised. The imagination of African unity has held power within the continent and in the African-American population of the USA which has also had to contend with slavery and subsequent official and unofficial expressions of racism. Authors such as W.E.B. Du Bois, C.L.R. James, Aimé Césaire and Léopold Senghor helped to bring anti-colonialism and ideas about African unity to a global audience. The international linkages inherent to pan-Africanism were expressed in the writing of US civil rights leader Malcom X and through the music of Bob Marley.

'Arab nationalism' has similarly sought to create an Arab identity but this runs alongside a vague sense of cultural belonging to a political belief in the possibility of establishing an Arab state. In the 1950s and 1960s a number of political movements came to openly support the idea of Arab nationalism, perhaps most prominently Egyptian leader Gamal Abdel Nassar, and the concept gained particular power in opposition to the establishment of the state of Israel in Palestine in 1948.

NEW WORLD ORDERS

Said has argued that the structure of difference between the Occident and Orient remained constant throughout those centuries he examined. Some have critiqued him for the lack of historical awareness that this exhibits, although, as we shall see in future chapters, there are many similarities between colonial and post-colonial cultures. Shohat and Stam (1994) have reworked Said's theory further to argue that in the middle of the twentieth century we witnessed a significant shift in the nature of the Occident as the USA rose to hegemonic dominance, having unparalleled influence on the political shape of the world but also in terms of cultural production.

Orders of Orientalism were reconfigured as modernisation theory and development. There have been various different variants of this, but most famously coming from Rostow's (1960) book *Stages of Economic Growth: A Non-Communist Manifesto*. The title of the book makes clear the new context for the post-colonial world order – the Cold War. Rostow's vision was of development which was driven by capitalism rather than communism, and thus it provided an alternative ideology to that followed by the Soviet Union and its allies. This was seen as a new, positive, post-colonial world order, one in which the differences between rich and poor would decrease and every country could make good. As we can see from the opening quote for this chapter, it was a period of great optimism. US President John F. Kennedy confidently proclaimed the 1960s the 'Decade of Development', and insisted that within ten years the differences between levels of economic development between the west and elsewhere could be overcome.

This then was an image of a truly post-colonial world order. The USA had been such a strong advocate of de-colonisation that the image it presented of the new US-centred world order was one based around equality. Looking back at this period from the vantage point of hindsight, to us this now may seem incredibly naïve and perhaps rather cynical. Certainly the 'Decade of Development' was not quite what it seemed.

Firstly, despite its egalitarian protestations, the USA desired the end of colonialism primarily to destroy trade relations between colonisers and colonies. The US economy was growing rapidly, producing goods for an American consumer society. However, while north America was a large continent it remained a limited market, and to ensure that growth could continue the USA needed to ensure a larger market. Therefore, it wanted a global free market, unimpeded by these special relationships between countries and their former colonies which made US goods more expensive in these countries. At the end of the Second World War, the USA had given vast amounts of aid to Europe in the form of the Marshall Plan. This was the first ever example of development aid. There were arguments related to altruism, but the main driver behind this was to ensure that Europeans would be able to afford US goods. The economies of European countries had to be rebuilt so that US goods could be sold and, without dynamic markets in Europe, US industrial growth could not be maintained. Thus, some have argued that development programmes to the ex-colonies were similarly targeted at developing markets for US consumer goods. At the same time, a weak economy was seen to offer a situation that would make a country vulnerable to take-over by the USSR. American geopolitics therefore suggested that it was important to bolster the economies of post-colonial countries to ward off the dangers of communism.

Secondly, and despite an apparent break from the past, the worldview of development was still based on a hierarchical and patronising model of the world: that there were developed and developing nations. Rostow and others presented the world as if all its countries could be located along one linear path to development, with the USA in the present and other countries located somewhere behind but aspiring to achieve the same heights.

From our vantage point at the beginning of the twenty-first century, of course, we can see that such development has not happened. Instead, differences in wealth have become larger.

THE THREE WORLDS CONCEPT

Nevertheless, the post-colonial world represents the emergence of a new configuration of global geography which seems to make the Orientalist model more complex by replacing a binary geography with one based on three spaces – the first, second and third worlds. However, as Carl Pletsch (1981) has argued, this new world order is in fact still based on binaries. The first binary is between the first and second world:

1 The **First World**, led by the USA, is based around free market capitalism and political freedom in democracy. It is taken by western theorists to be the natural expression of modernity, in that things are presumed to follow their own course without any intervention.
2 The **Second World,** led by the USSR, offers an alternative model of development, one which western theorists view as being distorted by ideology, the intervention in the market by the state, and the control of freedom also by the state. The Second World was understood to be modern but it was contaminated with a mixture of ideology that prevented it from being efficient, rational or natural, thus drawing on a colonial binary of the rational/irrational.

Both of these modern worlds were in opposition to:

3 The **Third World**, which is what was leftover!

The world is still based around binaries as Pletsch's diagram demonstrates but these operate at two levels. As with Said, Pletsch's first binary recognises the division of the world into the 'traditional' and the 'modern', but he also includes a second tier of binaries whereby the modern is divided into 'communist' and 'free'. Modernisation theory simply links these: all countries should inevitably modernise but will follow either path if swayed by one or other spheres of the modern.

So, what is it that unites the Third World? As the subtitle of Rostow's (1960) work suggests, this concept is primarily geopolitical. The Third World is no more than a residual category of the unaligned objects of the competing

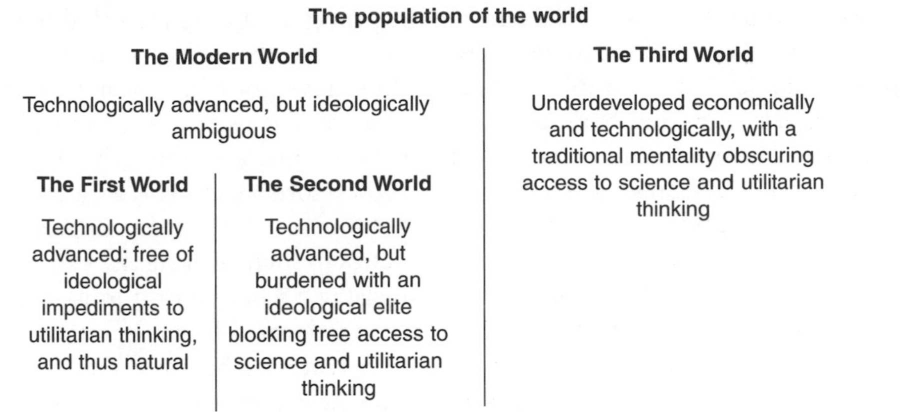

The population of the world

The Modern World		The Third World
Technologically advanced, but ideologically ambiguous		Underdeveloped economically and technologically, with a traditional mentality obscuring access to science and utilitarian thinking
The First World	**The Second World**	
Technologically advanced; free of ideological impediments to utilitarian thinking, and thus natural	Technologically advanced, but burdened with an ideological elite blocking free access to science and utilitarian thinking	

Figure 4.1 Pletsch's Three Worlds concept

policies of the first two worlds. This residual-ness was challenged by the Bandung Conference which attempted to propose a third way for non-aligned countries to avoid either form of western modernisation – a positive sense of an alternative to the west – but this was a very loose group of states that produced a rather superficial sense of identity. Some have indeed sought to reclaim the 'Third World' as a positive term, referring to the political stance of the alternative 'third way' (neither US-capitalism nor Soviet-socialism), and it is still sometimes used today as a political position, notwithstanding the end of the Cold War which had heralded the term originally.

Like Said, Pletsch sees this imagined geography as stretching beyond the confines of international politics or other political representations of the world. It is not just a theory about geopolitical relations between the USA and the USSR. He claims that these binaries also run through academic relations so that there is a post-colonial division of labour that differs from that which exists under colonialism. Again, we see Pletsch's links to Said's work. During colonialism and even before it, Orientalists were experts in certain systematic aspects of the Orient – language or literature or politics or art. Now, Pletsch argues, there is a clearer division between disciplines.

Those who studied the Third World were anthropologists and they only studied the Third World once the effects of modernisation had taken hold. Traditionally, anthropologists are area specialists. In their training, there is an important rite of passage where fieldwork must be undertaken in as strange or different a society as possible, so that the anthropologist can divest him or her self of their preconceptions and fully enter into the life of the subjects of the research. When they return they must write ethnographies – detailed descriptions and explanations of the life of those in the Third World – to make sense to westerners of a random set of instances of otherness. Traditionally, anthropology is a discipline that accumulates its knowledge in case studies rather than theoretical propositions, trying to make sense of the world in its pristine state to understand the processes through which different groups of cultural practices emerge. For some anthropologists, it is only possible to do this in 'primitive' societies that have not had lots of contact with other groups and have not developed or modernised – it is generally held that these groups have 'purer' cultural structures as a result. Those cultures with lots of change and modern effects therefore begin to be 'weakened' or 'diluted' as a result. The Third World is of interest as a set of examples of different cultures, studies of individual societies in their own right.

Pletsch also considered those who studied the Second World to be area specialists but their focus was on the distortions created by the communist ideology. Pletsch was writing during the Cold War, and he noted that those who studied states in the communist or socialist world were doing so in order to understand the effects that communism or socialism had on the 'normal' operation of an economy or society, rather than examining the countries or societies for themselves.

Finally, those who study the First World were assumed to be more properly systematic. These academics sought to create models, theories and understandings of the variety of processes making up contemporary society. In other words, they examined a economic processes or came up with theories of political behaviour or social formations. These are not studies of *First World* economics, but of economics, not of *First World* political theory but of political theory, not of *First World* social processes but of social processes. Pletsch argues that it was in the First World that academics worked to discover 'natural' laws of human behaviour. This was impossible in the Second World because of the distortions created by ideology and similarly in the Third World because of a lack of development. It was assumed that there were no effects of ideology in the First World, and that it was therefore uniformly modern. Thus, theories were generated here and proclaimed as universal. Just as with the production of colonial knowledge then, the west, and now the First World, was seen as the norm, the natural, the way it should happen. Everything else and everywhere else was regarded as a deviation.

Of course, the Cold War has now ended. There is now also an increasing emphasis on breaking down disciplinary boundaries. Could this be a geopolitical effect?

Further reading

On the emergence of a new world order

Pletsch, C. (1981) 'The three worlds, or the division of social scientific labor, circa 1950–1975', *Comparative Studies in Society and History*, 23 (4): 565–590.

On (post-) colonial nationalist identity

Chatterjee, P. (1986) *Nationalist Thought and the Colonial World: A Derivative Discourse*. Minneapolis: University of Minnesota.

Chatterjee, P. (1993) *The Nation and Its Fragments*. Princeton, NJ: Princeton University Press. (See especially the chapter 'Whose imagined community?'.)

Fanon, F. (1963) 'National Culture', in *The Wretched of the Earth*. New York: Grove Weidenfeld pp. 206–48.

On other alternative (geo)political formations

The website for OSPAAAL at http://www.ospaaal.com/

Young, R. (2003) *A Very Short Introduction to Postcolonialism*. Oxford: Oxford University Press.

5

COKE OR MECCA-COLA?
GLOBALISATION AND CULTURAL
IMPERIALISM

Figure 5.1   *The Independent*, 26 November 2003; Andrew Clennell

A group of British Muslim businessmen launched a campaign yesterday to tackle one of America's corporate giants. Thousands of cans of Mecca-Cola were handed out at mosques in Birmingham and Regent's Park in London, before a full-scale launch of the drink next January.

It all began last year at the whim of a 10-year-old. A French entrepreneur, Tawfik Mathloufi, asked his son to give up drinking Coke because of its American corporate association. His son agreed, but only if an alternative was provided.

M. Mathloufi saw a gap in the market and has had success with Mecca-Cola in France and the Middle East. His company gives 10 per cent of proceeds to Islamic Relief, which funds Palestinian charities, including an orphanage.

When Rashad Yaqoob, 31, a Yorkshire-born lawyer and investment banker, saw M. Mathloufi interviewed on the BBC, he tracked him down and offered to help him with legal advice and setting up the brand in Britain.

Now Mr Yaqoob aims to capture 5 per cent of the cola market in this country. Worldwide, he said, the company was selling 'about 50 million litres a month'. Most of that is in the Middle East but there is a significant market in France, where sales are at 800,000 litres a month. 'There's a very strong anti-American sentiment', Mr Yaqoob said. 'Coca-Cola represents the excess of corporate America.

'We wanted to give a bloody nose directly to the number one corporation that represents corporate America because corporate America represents Bush and Bush represents neo-conservatism'.

[…]

Mecca-Cola has previously given out thousands of cans at Stop the War rallies.

[…]

In keeping with Muslim ideals, the Mecca-Cola can states: 'Please do not mix with alcohol'.

In this chapter, we will look at the ways in which other cultures are caught up in **globalisation**: the effect of capitalism rendering difference as a set of products to be consumed, and at the same time, the effects of global culture on other cultures around the world. We are all used to hearing people bemoan the fact that places everywhere are beginning to look the same: you can buy Coke, McDonald's burgers and Starbucks' coffee in increasing numbers of countries; satellite TV beams music and fashions to teenagers worldwide; the internet now allows for instantaneous global communications whether through news pages, blogs or YouTube. Such has been the power of American culture, that cultural distinctiveness around the globe is being eroded, this argument continues. **Cultural imperialism** – the spread of global (or, in some versions, American) ideas and cultures – has steadily come to replace the formal imperialism and colonialism of the past. However, following the opening example, we will now examine these claims, and will find that while there are clear markets exploiting cultural difference, and while the power of global/American culture is unrivalled, cultural imperialism does not go uncontested: alternative expressions do persist, and there are still important cultural geographies existing in the post-colonial world. We will start with a consideration of how other cultures are consumed in post-colonial capital-ism, and then examine the nature of contemporary cultural imperialism.

CULTURAL IMPERIALISM AND THE 'SALVAGE PARADIGM'

Some have defined cultural imperialism as the use of political and economic power to spread the values, systems, ideas, and institutional forms of a foreign culture on a native culture. This spreading of cultural traits, from more powerful groups to less powerful groups, is a form of domination which accompanies and supports political or economic domination. Such cultural domination may be necessary for securing the hegemony neces-sary for other forms of domination. For instance, many would argue that the spreading of consumer culture through global brands such as McDonald's, Starbucks, Nike and MTV, is part of the power of America as the global hegemon, but is also central to developing a global culture that is accepting of US dominance. Talking of the success of *Rambo* and Chuck Norris films in Asia, Pico Iyer (1988: 5) makes the following observation:

> When [the US author] William Broyles returned to his old battlegrounds in Vietnam in 1984, he found the locals jiving along to 'Born in the U.S.A.,' Bruce Springsteen's anthem for the dis-enfranchised Vietnam vet, and greeting him with cries of 'America Number One!' 'America,' concluded Broyles, 'is going to be much more difficult to defeat in this battle than we were in the others. Our clothes, our language, our movies and our music – our way of life – are far more powerful than our bombs.'

Because the terms culture and imperialism are both very difficult to define, any definition of cultural imperialism will be the subject of a variety of different debates. The very possi-bility of cultural imperialism is premised on the idea not of culture (as in the old fashioned sense of civilisation or high culture), but the conception of a *plurality* of cultures: that there are many different ways of life all equal in their right to exist autonomously, a plurality of authentic or indigenous cultures which somehow properly belong each to their own area. Globalising or homogenising forces then threaten the integrity of these local cultures.

These individual, autonomous cultures are usually seen as 'traditional' and relatively untouched by capitalism as a modernising and homogenising cultural force. It is the processes of colonialism, globalisation and capitalism that are corrupting. With the spread of multinational capitalism it is assumed that this culture of consumerism will begin to overwhelm many aspects of native cultures. This process is therefore seen as one of loss. This perspective has often been adopted by anthropologists and is sometimes known as 'the salvage paradigm'; the desire to salvage what are mistakenly assumed to be 'timeless cultures' that are seen to be threatened by the onslaught of global culture.

CONSUMING THE OTHER

We will start this chapter with a consideration of how the 'other' has become an important commodity in post-colonial global culture in terms of selling difference in products from fashion to film, and by selling places in tourism. These seem like fairly harmless examples, but Gregory (2004: 10) explains that:

> This is not a harmless, still less a trivial pursuit, because its nostalgia works as a sort of cultural cryonics. Other cultures are fixed and frozen, often as a set of fetishes, and then brought back to life through metropolitan circuits of consumption.

Just as it was important to understand how colonialism was popularised and how ordinary people came to learn about the world, it is also important to understand popular understandings of the world of post-colonialism. It is impossible to do this without considering the role of the media which – as the name suggests – mediate all of our understandings of and interactions with the rest of the world. We will first consider the role of film and fashion in the reproduction of images of other places and people, before discussing the role of more educative media such as *National Geographic*. We will finish up by looking at the role of tourism.

Film

Both the images of colonialism and the consumption of otherness are central to contemporary culture. Britain in the 1980s and into the 1990s witnessed an unparalleled revival in films about the empire and colonialism. This coincided with the Thatcher government which used the rhetoric of Victorian values and even of empire in its discussion of Britain at the time. In the 1980s alone various films were produced that created a romanticised account of the colonial experience, most especially the experience of colonialism in India, such as *The Jewel in the Crown, The Far Pavillions, Staying On*, A *Passage to India, Gandhi, Heat and Dust* and *Kim.*

In an essay published in 1984, the author Salman Rushdie described the rise in interest in colonial history at that time:

> Anyone who has switched on the television set, been to the cinema or entered a bookshop in the last few months will be aware that the British Raj, after three and a half decades in retirement, has been making a sort of comeback. After the big-budget fantasy double-bill of *Gandhi* and *Octopussy* we have had the black-face minstrel-show of *The Far Pavilions* in its TV serial incarnation, and immediately afterwards the overpraised *Jewel in the Crown*. I should also include the alleged 'documentary' about Subhas Chandra Bose, Granada Television's *War of the Springing Tiger*, which, in the finest traditions of journalistic impartiality, described India's second-most-revered independence leader as a 'clown'. And lest we begin to console ourselves that the painful experiences are coming to an end, we are reminded that David Lean's film of *A Passage to India* is in the offing. I remember seeing an interview with Mr Lean in *The Times*, in which he explained his reasons for wishing to make a film of Forster's novel. 'I haven't seen

Dickie Attenborough's *Gandhi* yet,' he said, 'but as far as I'm aware, nobody has yet succeeded in putting India on the screen.' The Indian film industry, from Satyajit Ray to Mr N. T. Rama Rao, will no doubt feel suitably humbled by the great man's opinions. (Rushdie, 1991: 87)

Whether celebrating Gandhi, or more generally the romanticised history of the British in India, these productions remained firmly locked within elements of colonial thinking. Rushdie argued that Britain was in danger of entering a condition of psychosis, strutting and posturing like a key player on the world stage while the country's power was actually diminishing. The image that these films gave was that the British and Indians actually understood each other jolly well, and that the end of the empire was a sort of gentleman's agreement between old pals at their Club. This was true even of more revisionist accounts. David Lean's version of Foster's *A Passage to India* was not as critical as Foster had intended and toned down the more anticolonial notes in the name of 'balance'. The film was still generally positive, in that despite the flaws, violence and meanness of empire it was still a fundamentally glamorous affair.

Richard Attenborough's film of the life of Gandhi seemed to offer the opportunity for a more critical account of British colonialism. However, Rushdie argued that the line of criticism was muted: Gandhi was a man who irritated the British immensely but was safely dead. Once again, Rushdie was critical of the way in which the story was told:

1 It was driven by an exotic impulse which transcended the mere historical details, expressing a western wish to see India as a fountainhead of spiritual-mystical wisdom.
2 The narrative fulfilled a Christian longing for a 'leader' dedicated to the ideals of poverty and simplicity, a man who is too good for this world and is therefore sacrificed on the altars of history (paralleling the life of Christian Britain's hero Jesus).
3 And finally, the film expressed a liberal-conservative political desire to hear it said that revolutions can and should be made purely by submission, self-sacrifice and non-violence *alone*, and also that the British being civilised they thereby realised such moral reasoning was a good argument for them to withdraw!

Artistic selection changed the nature of certain historical developments by putting a different spin on them. For example, the film narrates the Amritsar massacre in the Punjab, when hundreds of Indians engaged in peaceful protests were shot at and killed by British soldiers. Both the massacre and the unrepentant scenes of the officer in charge, Dyer, were presented in the film as acts of a cruel and over-zealous individual who was immediately condemned by the Anglo-Indians. This individualising of blame thus absolved the colonial system and the British public of any complicity. But this was not the case. The British in the Punjab of 1919 were panicky and feared mutiny. As in the film's account, Dyer's court martial may have condemned him but

Figure 5.2   Indiana Jones as saviour

he was still given a hero's welcome on his return to England. There was sub-sequently an appeal fund raised for him and he thereby became a rich man on the proceeds.

There is also the issue of the film being an outside production 'speaking for' Indian history. The film was predominantly produced by, funded by, and principally acted by non-Indians. Indian history was being retold by the colonisers themselves. It could be argued that during the colonial period the British extracted value from the textiles, tea and other goods from India. With *Gandhi*, it was cultural resources that were being exploited – the film extracted value from Indian history and culture.

Other films have demonstrated the continuity of Orientalist categories. For instance, the hugely successful *Indiana Jones* films illustrate the triumph not only of the western individual hero, but also of western knowledge and leadership. Indy rescues artefacts from the colonised world for the benefit of science and civilisation and to recognise the value of artefacts that the indigenous people otherwise do not value – in the second film, it takes Indy to save the artefact, the blonde female lead and the village. The *Indiana Jones* films continue the 'Boy's Own' style of adventure discussed earlier which pro-vided a narrative of colonial adventure with charm. The exotic locations of the film indicate none of the complex politics and resistance of colonial period. Instead, the Orient is both demonised and infantalised. Often the 'good' natives are children under Indy's reluctant guidance (hinting back to the paternalism of the 'white mans' burden').

The *Indiana Jones* films, and the various others discussed by Rushdie, were mostly made in the 1980s. Have the ways in which the Orient has been represented in mainstream film and TV changed in any significant ways since that time?

Fashion

As Said argued, Orientalist images can be found throughout different cultural representations and productions, including, of course, the more visual forms of culture. Advertising, like film, is a very visual medium and in its aim of selling a product, it very consciously uses images. Although it is 'just an advert', it is important to remember that a lot of thought has been given to how each advert will work, what will be the most effective way of communicating a particular message, who the target customers are, and what characteristics and beliefs these people will hold. Fashion and advertising are highly dependent on familiar and easy-to-recognise imagery, and here too we find a particular image of 'the rest'.

Fashion advertising deals with colonialism and its effects in a number of ways. Perhaps most common is a depoliticised rewriting of history. In such cases, adverts present an aestheticised version of the colonial period which selects visual elements from various times and places to create a romanticised version of reality. For example, Ralph Lauren's imagery in his *Sahara* range clearly draws upon filmic versions of empire, hinting at *Out of Africa* and an aesthetic vision of crisp whiteness in the desert.

Another example comes from the high street. The US store, Banana Republic, started as a catalogue company but was bought by Gap in 1983. The name 'Banana Republic' is usually used contemptuously to refer to postcolonial Latin America countries, suggesting a lack of power and incompetent rule. But the imagery in these stores referred to colonialism more generally. In the 1980s and 1990s stores concentrated on safari and travel wear and were laid out like colonial general stores.

The *Petermans* catalogue similarly drew on such images of empire. The description of clothing for sale draws upon the traditions of travel writing:

> Lord Kitchener reconquered the Sudan wearing the 4-pocket khaki jacket that evolved into the modern safari jacket ... [we offer] the authentic bush jacket for adventurers with a low tolerance for the ersatz.

This is an uncritical celebration of European colonialism focusing on movie versions of the 'empire' look, without any reflection on the nature of the experience or the power relations involved. On one level, however, perhaps it is unreasonable to critique Petermans and Banana Republic for this message. Why should they draw attention to the negative parts of history? Is this the job of companies selling clothes? It would probably not increase sales if the advertising is full of challenging and unsettling images. Nevertheless, it is important to understand why particular images are so often repeated in advertising, images which reinforce messages received from other media, to tell us that this is what the world is like. Because advertising is so consciously and meticulously thought through, the fact that these particular images are used is evidence that they are perceived by advertisers to work and is further proof that Said was correct in pointing out the persistence of images of colonialism and difference throughout contemporary culture.

Figure 5.3   This image was used as an advert in the Sunday magazine of the *New York Times* in 1991

Take a look at the picture making up Figure 5.3:

- **What symbolism is in this image?**
- **Which Orientalist discourses are drawn upon here?**
- **What might a feminist response be to this image?**
- **What is the role of humour here?**

Look at other fashion advertising that is set in countries outside of the west and think about the discourses drawn on. How much of a break do they make with Orientalist traditions?

Finally, we should turn to Benetton. Perhaps here we have a set of advertising images that do challenge our inherited wisdom about geographies of us and them. Known for its controversial advertising, Benetton has played with images of the post-colonial world order and other taken-for-granted values. However, once again its use of controversy only works because the company knows it can expect certain Orientalist views of the world to be held by most of the people who view the adverts. The store's slogan 'United Colours of Benetton' is drawing upon what Roland Barthes called the 'family of man' (1956). This superficial argument suggests that we are now all equal, but varied and colourful, and celebrating a post-colonial 'different but equal' message. (This of course misses the power relations infusing the global fashion industry, where workers in the Third World in sweatshops on meagre wages create goods that are sold at high prices in the west.)

There are numerous images from Benetton campaigns that could be examined, but we will look at only one:

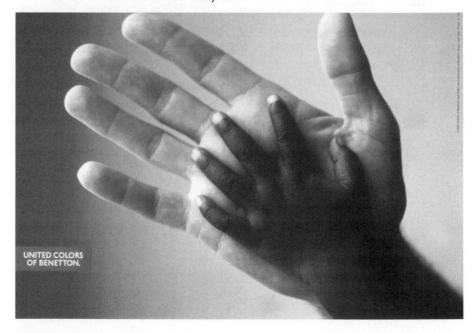

Figure 5.4 Benetton advert

In some respects this is a positive and innocent image where an adult hand is offering friendship and perhaps protection to the child's. But, the white hand is bigger and is active: it protects the child's hand, echoing colonial paternalism justifying European rule. Think back to the binaries running through colonial knowledge: the white hand here is male, adult and active, the black hand is childish and passive. Why aren't the colours reversed in this image? Benetton talk about the 'United Colours' of the world, but binaries of black and white still persist in a very particular hierarchy. What would happen if we did swap the colours here? We can see how shocking it can be when the image is reversed, as with a campaign run by the charity Drop the Debt.

Figure 5.5   Drop the Debt advert

Popular anthropology: *National Geographic*

If we think about where we have learned about how the rest of the world looks, many of us would probably point to *National Geographic* magazine. It has proved very influential. Read by around 37 million people worldwide, in the USA it is the third most popular journal after the *TV Guide* and *Reader's Digest*. Unlike the more frivolous images used in films and advertising, *National Geographic* claims to present truthful images of the world to its US readership. Nevertheless, if Said was correct about the on-going relevance of his arguments in *Orientalism*, we should see the same kinds of representations running through this journal as we have found previously in advertising and movies.

The *National Geographic* association has links with academic geography; it is widely regarded as educational and so has great scientific legitimacy.

However, despite these claims, it is a specifically American view of the world, which focuses a large amount of attention on the post-colonial countries of the majority world. Drawing on traditions of geography and anthropology, it tells Americans about the rest of the world. Some would argue that the *National Geographic's* photos are among the most powerful images of 'others' for Americans.

Just as with the Orientalist paintings we discussed in an earlier chapter, the *National Geographic* creates a 'reality effect' through the use of detailed photography in a respected magazine. With the increasing use of digital manipulation techniques we are perhaps now more wary of photographic images, but for most of the period of their use they have been regarded as an authentic or natural record. To enhance the reality effect, most of the photos used in *National Geographic* are simple, and do not use complex or artistic techniques.

Photos are also seen as a democratic form of representation in that everyone can understand a photo. There is a common sense as to how they are understood. The richness of exotic photos presents an enticing world beyond the suburban reality of most of its readers' daily lives, suggesting that in other parts of the world things are marvellously different.

Catherine Lutz and Jane Collins are two anthropologists who have carried out an in-depth study of *National Geographic*. They argue that we cannot see photographs as such an innocent representation:

> The photograph can be seen as a cultural artefact because its makers and readers look at the world with an eye that is not universal or natural but tutored. (Lutz and Collins, 1993: xiii)

The view is white, educated and middle class. Lutz and Collins also see the vision as gendered male because of associations with exploration and adventure. However, this view of the world is projected as being national and even universal.

The magazine's upbeat style asks readers to look at the rest of the world from the vantage point of the world's most powerful nation – which has the effect of reinforcing American values. Lutz and Collins found the existence of a number of themes running through the magazine's articles. For instance, they saw the repeated use of photographs of family-like groups. They argued that this creates a sense of order and logic that validated US family arrangements, regardless of how the people photographed had come together or what their real social relations might be. Nuclear families are therefore naturalised. We also return to the image of the 'family of man', showing the existence of fundamental human values and highlighting the fact that difference exists only at the level of superficial (but colourful) rituals and clothing styles. However, the idea of similarities between peoples was very much on western terms – western ideas of 'normal' family relations are asserted as human norms – and beyond this again, as with Benetton, this projection of similarity is deceptive. For example, it would seem that while it was acceptable for non-white peoples to be represented as naked the same standards did not hold for white people, and there have been accusations that light, naked skin had to be darkened to be considered respectable. Finally, any comparisons were firmly controlled. No images were included after the end of colonialism that drew attention to previous colonial relations.

Lutz and Collins's analysis of *National Geographic* between 1950 and 1986 suggests that it idealised and rendered exotic the Third World's people and downplayed violence and poverty. From their work they put forward six underlying issues to this representation:

1   **It concentrates on unchanging parts of cultures.**
2   **It looks at all groups as if they were on a trajectory from tradition to modernity.**
3   **It sees all peoples as being fundamentally the same underneath the superficial difference of culture.**
4   **It deflects attention away from power differences and exploitation, as they say:**

> images of westerners were politely removed from colonial and neo-colonial contexts, thereby avoiding uncomfortable questions about the nature of that presence, obscuring the contexts and difficulties of the photographic encounter, and creating a vision of the cultures in question as hermetically sealed worlds. (Lutz and Collins, 1993: 39)

Almost one third of pictures analysed had one or more people smiling.
5   **Most articles avoided indicating levels of wealth.**
6   **The majority of articles were set in rural areas. When the magazine showed modernity and progress, this was often presented as co-existing without conflict with traditional culture, and the humour of this juxtaposition is often highlighted (for instance 'backward' natives using video technology).**

Go and find a copy of the *National Geographic* and see whether you agree with Lutz and Collins's analysis. What image does the magazine present of the world as a whole? Who and what is included, and who and what are excluded?

On the one hand, we could argue that *National Geographic* was dignifying different peoples through the attention given to the details of their culture, but on the other hand, we could say that the magazine simply rendered difference as nothing more than aesthetics: the exotic as something which sells. Much of this difference was identified through a narrative of dress and clothing, suggesting local difference at a level of aesthetics and timelessness. This may work against a more profound understanding of differences between people emerging from deeper economic, social and political processes. The magazine certainly drew attention away from any interpretation which would suggest that there were connections between places, particularly in terms of (neo)colonial relations which would mean that the wealth of one part of the world was the result of others being kept in poverty. The only relationship between people which mattered was our belonging to the 'family of man'.

Tourism

Tourism is probably the most important mediation of otherness in terms of our sense of experience of difference. International tourism is big business and whether we have actually travelled ourselves or just looked at the relevant advertising, our impression of other places will have been significantly mediated by the tourist industry.

Before the nineteenth century there was no tourism, only elite travel. In the case of Britain, wealthy families would organise grand tours for their sons to go abroad and learn about the civilised world, to be improved by the great historical sites of continental Europe. Very few, however, went beyond Europe as there was no provision for tourists and there was significant danger even if the lands had been colonised. But the nineteenth century did still witness the rise of package tours (introduced by Thomas Cook). These organised trips took the burden of planning from the individual traveller. With the rise of middle-class values, particularly once the twentieth century had arrived, travel increased still further, but it was not until after the Second World War that there was a rise in mass tourism – when it truly became within the reach of the middle classes and increasingly working-class people as well. Such travellers tended to be more dependent on package tours than the self-organised (and more expensive) tours of the elite. The increased affluence, leisure time and travel facilities of the late twentieth century have thus seen tourism boom into an industry worth more than $2.5 trillion. And people are also travelling further. In the mid-1970s, only 8 per cent of North American and European tourists travelled to Third World destinations. This figure is now well over 20 per cent.

So what have been the effects of tourism on host countries, and visitors to them?

International agencies together with European, American and Japanese bankers have supported the development of Third World tourism. This industry exists to turn poverty into a good in the market-place for foreign currency, as what visitors want here is an absence of western commodities and exotic difference coupled with cheap wages. It is a valuable industry, but only if visitors get what they expect. This has in some places led to repressive political policies to stop political unrest and violence. Many countries, such as Egypt and Jamaica, have very evident tourist police in order to assure visitors' safety.

Tourism has become a major industry and offers new possibilities for those ex-colonies whose economies had been shaped during the colonial period. One of the major problems of colonial production was that the colonised countries tended to be developed by colonisers to produce one crop only (e.g., sugar, cotton, coconuts or minerals). These countries were then overdependent upon that one good, so that when market demand dropped for that commodity there were severe economic repercussions. Tourism posed an alternative to this one crop dependency. For example, in the mid-1980s tourism replaced sugar as the top foreign exchange earner in the Dominican Republic and this was also the case for bauxite in Jamaica. Countries as diverse as Puerto Rico, Haiti, Nepal, Gambia and Mexico now spend millions in development funds for developing tourism. Even socialist countries such as Cuba and Vietnam are encouraging tourism. Is there a danger of these countries becoming overdependent upon the tourist industry now, just as they may have been on a single commodity before?

Why is the Third World such a draw for tourists? Just as we have seen in terms of films and advertising, tourism aestheticises otherness: to be poor in the late twentieth and early twenty-first centuries society is presented in tourism terminology as being 'unspoilt'. Furthermore, while tourists want difference, it is *visual* difference and not real difference that they are after. Most tourists do not expect to be daring, to learn a language or adapt to local customs. Local currency may be their only form of adaptation.

This suggests that, contrary to the imagery of tourism, to an extent local cultures will have to be adapted to western standards. This can be seen as the domestication of different cultures into western-friendly forms, to conform to western expectations. Critics suggest that different cultures become like theme parks where visitors can see (and, importantly, photograph) different things, but are then able to return to their hotels, eat westernised food (or local food selectively adapted to a western palate) and live with western-levels of comfort. Again, the exotic and different are rendered commodities to be purchased by western consumers.

Package tours insulate tourists even further from the locals. Here tourists may only come into contact with local people under carefully controlled conditions. For instance, tour agencies might set up a time for a performance of local cultural rituals. Tourists then come and watch, photograph, and leave. This is truly an interesting phenomenon – what we might consider to be a paradox of tourism – which is that as an industry it depends upon difference (people want to go and see something different) and yet the process of tourism fosters global integration, destroying the very thing tourists come for.

The construction of different cultures and travel itself in the western imagination can be seen through advertising which draws on a number of different narratives and discourses.

The experience of travel is thought by many to transform a person through this exposure to difference. This idea dates from eighteenth century Romanticism, which longed to escape civilisation to 'return' to an unchanging nature or primitive society. Thus tourists seek authentic cultures which have not been overlain by western culture. Locals create such authenticity for the tourist trade (namely, what they think tourists will want to see). Paradoxically tourists then come to see the local culture but instead see a *performance* of the local culture, because the reality of local life involves a degree of modernisation. At one extreme, local people will put on traditional costume, perform, and then go home and change into jeans. In the Amazon, native people will lay on performances of 'traditional' identity for visitors several times a day to fit into cruise schedules because this activity is what is perceived to be real. The reality of the performers' day-to-day life is itself too mundane for tourists. Look at travel adverts and you will see phrases like 'life that has remained unchanged for centuries', 'life at a different pace', and 'leave mobile phones and the internet behind, and step into a simpler way of life'.

Tourism can also be linked to the historical drive to find a Paradise or Eden. The European search for the location of Paradise moved as people travelled across the world in the Age of Exploration. Initially it was believed to be in modern-day Ethiopia, and then when that land was discovered, Paradise was always just beyond what was the known world. In the eighteenth century many described the South Seas in this way because they felt the natives had ready food (just dropping off the trees), a warm climate, and what were perceived (by the male sailors) as ideal social relations, where everyone seemed friendly, there was no violence and plenty of sex. 'Paradise' is the most used word in travel adverts, so this might suggest these eighteenth century meanings are still important to the reason for (and locations of) tourism.

MODERN DAY ORIENTALIST TRAVEL

The December 2006 issue of the luxury travel magazine, *Condé Nast Traveller*, included a 'traveller promotion' on India which listed a number of the themes mentioned above. An advertisement in this section selling train travel in Maharashtra makes the following claims for the holiday experience:

> On the Deccan Odyssey, you travel through mountains, forests, beaches and time itself. Rediscover Nature. Then discover the history of civilization: Battle-scarred forts; peace-inducing temples, ashrams and churches; elevating Buddhist caves. In seven days, this one-of-a-kind train takes you from Mumbai to Jaigadh, Ganapatiphule, Sindhudurg, Ratnagiri, Goa, Pune and Aurangabad. Actually, it transports you to another realm. And the wonders that await you every time you alight from the train are matched only by the splendours on board: A multicuisine restaurant fittingly called the Peshwa; the Mumbai Hi Bar and Lounge; a gymnasium; the Deccan Spa with its rejuvenating massages ... Religion. History. Art. Spiritualism. Whatever be your destination, the Deccan Odyssey will take you there.

There are a number of problems with the post-colonial rise of tourism. Firstly, is its hastening of cultural imperialism which may leave changing value systems and expectations through contact with western consumers.

Environmental impacts have also been documented, especially in terms of western consumers' expectations of clean linen, air-conditioned accommodation and convenient transport. And, frequently, tourist money gives little benefit to local economies as it is often channelled back to the west through western-owned tourist hotels.

There have been various suggestions of how these problems can be managed. Bhutan has decided to limit the number of people who can enter each year, and in many countries there have been attempts to develop low-impact or eco-tourist industries. While these do indeed minimise the cultural and economic impacts on host countries, they still have other impacts. By restricting the number of people who can visit a place (through direct restrictions or by developing low-density, low-impact accommodation) then the market will respond by raising prices. Is the response to the effects of mass tourism going to price out the masses? Will tourism become something that is open only to those with the most money, and if so, what might this mean for the tourist encounter, and for all of our geographical imaginings of people and places that are different from home? Perhaps we might want to challenge these critiques of tourism as being too extreme. Are there no benefits to tourism? We could argue that there are real benefits to western people experiencing difference (even if it is controlled by the tourist experience) or, in a very materialist way, that the financial rewards for poor countries are more important than any of the cultural drawbacks. Think about these questions in regard to your own experiences of travel – how open to difference do you think you have been?

GEOGRAPHIES OF CULTURAL IMPERIALISM

Despite the issues raised so far in this chapter, many people in the poor countries of the global south see modernity in the form of material prosperity as something worth having – and for many it is something that is greatly desired – even if some western values happen to accompany it. It could be argued that the critique of cultural homogenisation may be of more relevance to those in the west who want to keep the world a diverse and interesting place to visit on holiday, or to read about in *National Geographic*, or as a source of 'ethnic' things with which to decorate their homes. This raises the issue of our right to speak on behalf of others. Do we then get to decide what is to be presented and what is allowed to change, and who is allowed to benefit from modernism?

There is a danger here of treating others as cultural dupes who fail to recognise their own interests in a paternalistic gesture. In fact, many people in Third World countries *do* indeed worry about western secularisation corrupting their youth; nevertheless, they are anxious for the material benefits.

Figure 5.6 Anti-Coke graffiti in India

Images of global homogenisation often assume that everyone aspires to western/ American values and the domination of western cultural and material goods, and therefore will forget traditional values.

This threat of the moral disintegration of a culture is sometimes manipulated by ruling elites who wish to direct attention away from their own exploitative practices towards an external enemy. Cultural reproduction has thus become highly politicised. Where influences are seen as coming from the outside there is sometimes an increased awareness of and appreciation for the local. The threat of global culture and homogenisation can lead to the revitalisation of local cultures and the invention and re-establishment of traditions. This can usefully stimulate debate over cultural values and critiques of the universalisation of western norms, which in some cases can lead to a rejection of western values and occasionally a strong reaction against the west for its perceived decadence. Evidence cited for this decadence includes its sexual mores and high divorce rate (levelled especially by Islamic cultures). Sometimes the west is demonised for its evils as evidenced in the violence of western society. Some societies wish to strengthen their own values in the face of values exported from the west.

Basically then there are two responses to globalisation. Firstly, there is an often highly self-conscious management (and sometimes, defence) of local cultures. Secondly, there is also an increasing awareness of the celebration of cultural syncretism and creolisation that is often discussed in terms of hybridity. These have also been termed 'McWorld vs. Jihad' by Benjamin Barber, recognising the cultural values attached to each and also showing, as with the example we started this chapter with, that globalisation and opposition to it are enmeshed within.

> Just beyond the horizon of current events lie two possible political futures – both bleak, neither democratic. The first is a retribalization of large swaths of humankind by war and bloodshed: a threatened Lebanonization of national states in which culture is pitted against culture, people against people, tribe against tribe – a Jihad in the name of a hundred narrowly conceived faiths against every kind of interdependence, every kind of artificial social cooperation and civic mutuality. The second is being borne in on us by the onrush of economic and ecological forces that demand integration and uniformity and that mesmerize the world with fast music, fast computers, and fast food – with MTV, Macintosh, and McDonald's, pressing nations into one commercially homogeneous global network: one McWorld tied together by technology, ecology, communications, and commerce. The planet is falling precipitantly apart *AND* coming reluctantly together at the very same moment. (Barber, 1992)

Writing in a very similar vein in a newspaper article in the aftermath of the Al Qai'da attack on the Twin Towers, Naomi Klein (2001) noted that:

> In Afghanistan you can buy t-shirts bearing counterfeit Nike logos and glorifying bin Laden as 'the Great Mujahid of Islam'.

As geographers, we might also want to question whether just because we can identify a process as being global in its influence, whether this means that all parts of the globe will be affected in the same way, and thus whether global processes are really rendering the globe more and more similar. Geographer Doreen Massey has acknowledged that the globe is effectively shrinking as a result of the effects of globalisation, but this process is not evenly spread (see the box below on 'power-geometry'). The effects of globalisation are felt differently all over the world. Figure 5.7 shows the highly uneven geography of one of the things more often quoted as evidence of the destruction of geographical difference, the internet. We should not consider culture to be singular and unchanging. We can view the new cultures and practices that emerge as a result of cultural merging as part of normal cultural processes if we consider culture not as unchanging or timeless but as always being made and remade.

'POWER-GEOMETRY'

Doreen Massey is very critical of overly-simplistic views of globalisation that suggest the world is getting smaller, what has been called 'the annihilation of space by time' or 'time-space compression' by different theorists. She suggests that while this global shrinking has been experienced by some (wealthy, white, mostly male), the world has in some cases got

(Cont'd)

larger and more difficult to traverse for others (poor, non-white, 'illegal', often female). She proposes an alternative way of understanding the fluidities and enduring barriers that are characteristic of contemporary migration, starting with:

> ... the question of to what extent its current characterization represents very much a Western, colonizer's view. The sense of dislocation which so many writers on the subject apparently feel at the sight of a once well-known local street now lined with a succession of cultural imports – the pizzeria, the kebab house, the branch of the middle-eastern bank – must have been felt for centuries, though from a very different point of view, by colonized peoples all over the world as they watched the importation of, maybe even used, the products of, first, European colonization, maybe British (from new forms of transport to liver salts and custard powder); later US products, as they learned to eat wheat instead of rice or corn, to drink Coca-Cola, just as today we try out enchiladas.

> [...]

> Now, I want to make one simple point here, and that is about what one might call the *power-geometry* of it all; the power-geometry of time-space compression. For different social groups and different individuals are placed in very distinct ways in relation to these flows and interconnections.

> [...]

> In a sense, at the end of all the spectra are those who are both doing the moving and the communicating and who are in some way in a position of control in relation to it. There are the jet-setters, the ones sending and receiving the faxes and the e-mails, holding the international conference calls, the ones distributing the films, controlling the news, organizing the investments and the international currency transactions. These are the groups who are really, in a sense, in charge of time-space compression; who can effectively use it and turn it to advantage; whose power and influence it very definitely increases. On its more prosaic fringes this group probably includes a fair number of Western academics.

> But there are groups who, although doing a lot of physical moving, are not 'in charge' of the process in the same way. The refugees from El Salvador or Guatemala and the undocumented migrant workers from Michoacán in Mexico crowding into Tijuana to make perhaps a fatal dash for it across the border into the USA to grab a chance of a new life. Here the experience of movement, and indeed a confusing plurality of cultures, is very different. And there are those from India, Pakistan, Bangladesh and the Caribbean, who come halfway round the world only to get held up in an interrogation room at Heathrow.

> Or again, there are those who are simply on the receiving end of time-space compression. The pensioner in a bedsit in any inner city in this country, eating British working-class-style fish and chips from a Chinese take-away, watching a US film on a Japanese television, and not daring to go out after dark. Massey (1993: 59, 61–62)

Studies of the effects of cultural imperialism must certainly include the effects of television and cinema. There is a tendency to think of the media as being dominated by the USA but this is only partially true. Certain things do

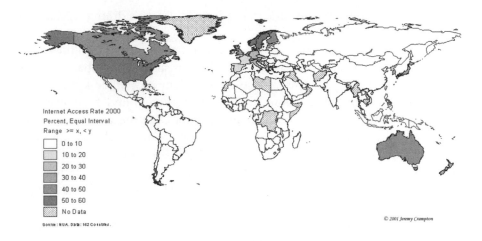

Figure 5.7   Uneven geography of Internet access rates, 2000

export well from Hollywood, especially violence. Media such as film and television have been an important part of the globalisation and homogenisation of cultures. However, they are also increasingly part of the resistance to such globalisation: think of the use of radio broadcasts in the Algerian fight for independence (evident in the film *Battle of Algiers*), or the use of the internet by resistance movements, with the most well known being the Zapatistas in Mexico, in an attempt to connect local struggle to global awareness through the institutions of the media.

Some are more apt to celebrate hybridity in their cultural systems. Bollywood films take Hindu themes and respect certain taboos (no kissing, but otherwise plenty of [non-sexual] contact), but with westernised filmic backdrops and westernised Hindu music. However, of all the types of television programming soap operas may be the most influential. Soap operas are an important area in which there is clearly no unilateral forcing of First World products on a passive Third World. In this form, global mass culture does not so much replace local culture as coexist with it. There are also powerful reverse currents in a number of Third World countries, such as Mexico, Brazil, India and Egypt. These countries not only dominate their own markets, they are also large exporters.

Soap operas are hugely popular with tens of millions of people around the world. Although North American soaps are broadcast widely around the world, these are outnumbered by the astounding global circulation of serials made in other cultures. British serials are sold to around 20 countries and Mexico's *Televisa* exports its serials to 59 countries, including the USA, but the world's most successful exporter of serial drama is Brazil's *TV-Globo*, the world's fourth largest television corporation. *TV Globo*'s dramas are seen in more than 100 countries.

In addition, there are various ways in which people respond to the stories they see on TV that show the complexity of the meanings of globalisation (as Massey argued). Soap operas, perhaps more than any other kind of programming, stimulate discussions about the characters and their everyday lives. Soaps are seen as examples of daily life. They can produce debates about characters and issues, even incorporating moral lessons. Cultural values are often the topic of these conversations (as the UK's BBC used to say of its soap *EastEnders*, 'Everyone's talking about it'). Such issues as generational conflicts and modern urban ideas are often the subject of such debate, brought on by the intense interest and an identification by viewers with the TV characters. Religious and ideological values are also often dramatically highlighted in these serialised narratives.

Television as a medium plays a somewhat different role in different cultures, depending upon the traditions of storytelling in that culture. During the religious period of Ramadan special soaps are made in Egypt. People are tired after a day of fasting and tend to spend time with their families, which make for ideal TV viewing times. The Indian TV versions of the Hindu religious texts, such as *The Mahabharata*, regularly attract incredibly high numbers of Indian viewers (up to 90 per cent).

RAMADAN TV

The holy month of Ramadan each year is an occasion in which Muslims fast between sunrise and sunset so as to be able to understand what it must feel like to go hungry and thirsty. This empathy with the world's poor should also result in the giving of alms and charity, another of Islam's central beliefs. However, this traditional religious festival has also resulted in new cultural forms, such as Ramadan TV.

> The sacred month is traditionally marked by long evenings at home when families break their fast, and this year Egypt has produced 50 TV series for the captive audience of Ramadan with Syria running a slow second at 45. (*Egyptian Mail*, 23 September 2006)

These special TV programmes are exported to countries throughout the Middle East and the gulf states. Although produced in the style of soap operas, these also have religious and moral messages. Amongst the 2006 line-up were *The Heathens*, about Islamic groups responsible for terrorist attacks in London and various Arab capitals, *Al-Hilali's Way*, about a politician whose corrupt past catches up with him, and others that were more directly interpretations of historical and religious dramas.

In India a film genre called mythologicals has been one of the most popular. Tradition is articulated through modern technology and a modern form of storytelling. Film and television versions of ancient stories take on flesh and blood and the impact is overwhelming. When Rama and Lord Krishna appear on screen the audience will often prostrate themselves. This devotional behaviour of the audience is strikingly different from typical viewing behaviour in the west (passive, knowing and increasingly ironic). This reaction to the television version of the Ramayan was a response that has a long history in India. The complete identification of the actor with a deity is well known in Hindu folk performance. The serialised narration of the Ramayana in India can attract a daily viewership of 60 to 100 million (roughly an eighth of India's population). There are Ramayan-related news stories in the newspapers everyday. In many homes watching it has become a religious ritual. Television sets are covered in garlands and decorated with sandalwood paste. Grandparents make their grandchildren bath before watching. Meals are postponed so that families are purified before watching Ramayan.

There is, then, a geography to the use of technologies. Here we can take a further example – of mobile phones which have now become synonymous with modernisation and globalisation. The spread of mobiles across the world could be regarded as part of the consumerism spreading through the middle classes throughout the globe. However, this is not straightforwardly the case. Although undoubtedly many phones are sold to business people wanting to stay in touch with the office, and to wealthy teenagers wanting to stay in touch with the latest trends (thus apparently spreading a version of modernism so hated by those who critique globalisation), the growth of mobile phone use can also been seen to protect and enhance place-specific and unique cultures too. For young Egyptians, like young people in other countries, having a mobile phone is important. However, for many (even middle-class people) the cost of calls prevents extensive use in a conventional sense. They are used in other ways to stay in touch. A friend can phone and the person receiving the call will see their name appear on screen and allow the ringtone to play for some time before hanging up. It is not necessary here to retrieve the call and pay for it. The caller will not expect this, instead the call is made to let the receiver know that his or her friend is thinking of them. In pastoral communities around the Tanzanian capital of Dar es Salaam, mobile phones are used by herders to keep in touch with market prices (a special service has been established to supply up-to-the-minute information) to ensure they are not cheated when they arrive in town to sell their animals. Camel drivers in the south eastern desert of Egypt are not able to pick up mobile signals, but they have been known to use satellite phones to keep in touch with the changing market prices for camels in the market town of Daraw, so they know when to get to market to achieve the highest prices.

RETURN TO THE 'SALVAGE PARADIGM'

The salvage paradigm sees history as linear and unchanging and always caught up with ideas about loss. It depends on a binary between the authentic and the false, where a number of authentic cultures around the world are destroyed by the inexorable spread of fake, insincere or inauthentic global culture. There is a real sense of particular cultures belonging in particular places, and others being 'out of place'. The salvage paradigm reflects a desire to rescue authenticity out of destructive historical change which it sees as a linear and non-repeatable process. It always involves the discussion of timeless cultures which are undergoing the impact of disruptive change associated with commodities, trade, the media, missionaries, ethnographies, tourists, and art markets. In short, it is being corrupted by modernity.

However, the examples discussed here suggest that there is a need to rethink 'authenticity' so that it does not simply mean pure and unchanging. It is possible for there to exist 'authentic cultures' based on hybridity and creativity. Furthermore, we need to think about why there is this sense of loss. Some have suggested that we are so concerned with salvaging the other because of our sense of what we have lost of our own past. This suggests that there is a fear from a western perspective that exotic cultures are undergoing fatal changes which will remove these vibrant parts of the global marketplace, but perhaps this also reflects the anxiety that it is western processes that have sounded the death knell for them.

Further reading

On globalisation and cultural imperialism

Appadurai, A. (1994) 'Disjuncture and difference in the global cultural economy', in P. Williams and L. Chrisman (eds), *Colonial Discourse and Postcolonial Theory*. New York: Columbia. pp. 324–39.

Barber, B. (1992) 'Jihad vs McWorld', *The Atlantic Magazine*, March. (www.theatlantic.com/politics/foreign/barberf.htm – his arguments have also been extended to a book of the same title.)

Iyer, P. (1988) *Video Night in Katmandu: And Other Reports from the Not-So-Far East*. New York: Vintage.

Massey, D. (1993) 'Power-geometry and a progressive sense of place' In J. Bird, B. Curtis, T. Putnam and G. Robertson (eds), *Mapping the Futures*. London: Routledge.

Said, E. (1993) *Culture and Imperialism*. New York: Knopf.

Shohat, E. and Stam R. (eds) (1994) *Unthinking Eurocentrism*. London: Routledge.

On film and fashion

Rushdie, S. (1991) *Imaginary Homelands: Essays and Criticism, 1981–1991*. New York: Granta.

Smith, P. (1988) 'Visiting the Banana Republic', In A. Ross (ed.), *Universal Abandon?* Minneapolis: Minnesota University Press. (Reprinted in *Social Text*, 7 (3).)

On National Geographic

Lutz, C. and Collins, J. (1993) *Reading National Geographic*. Chicago: University of Chicago University Press.

Rothenberg, T. (1994) 'Voyeurs of imperialism: the *National Geographic Magazine* before World War II', in A. Godlewska and N. Smith (eds), *Geography and Empire*. Oxford: Blackwell.

On tourism

Bruner, E. (1991) 'Transformation of self in tourism', *Annals of Tourism Research*, 18: 238–50.

MacCannell, D. (1976) *The Tourist: A New Theory of the Leisure Class*. New York: Schocken.

MacCannell, D. (1992) *Empty Meeting Grounds*. London: Routledge.

Try to get hold of a copy of Dennis O'Rourke's film *Cannibal Tours*. This charts the progress of an upmarket tour of Papua New Guinea from the perspective of the tourists and from the natives.

- What discourses are used by the tourists to describe Papua New Guinea?
- Why are the tourists there? What are they doing?
- Do the New Guineans understand?
- How are the tourists transformed?
- How are the New Guineans transformed?
- What are the images of the tourists and New Guineans produced by the film?

On soaps

Abu-Lughod, L. (1995) 'The objects of soap opera: Egyptian television and the cultural politics of modernity', in D. Miller (ed.), *Worlds Apart: Modernity Through the Prism of the Local*. London: Routledge. pp. 190–210.

Allen, R.C. (1995) *To Be Continued: Soap Operas Around the World*. London: Routledge.

PART III

POSTCOLONIALISMS

In the third and final section, we will think about postcolonialism as a positive political project which critiques western assumptions, stereotypes and ways of knowing and offers its own alternatives. One of the goals of postcolonialism is to include the voices of those excluded by dominant forms of knowing (because they have been considered to be 'native', 'unscientific', 'superstitious' or 'traditional' ways of knowing). Chapter 7 will address the postcolonial theory produced by university academics that challenges the high knowledges of western culture. While some of these address are quite abstract and often quite difficult, we will also come across the work of academics who want their ideas to make a real difference in terms of people's lives, and thus we will see the ways in which they have advocated change. In Chapter 8 we will move into the wider realm of culture to see how novelists, film-makers and musicians have sought to challenge western cultural dominance and to find ways of representing other experiences, belongings and identities through their work. Finally, in Chapter 9, and by way of a conclusion, we will consider both the limitations of postcolonialism and the ways in which it can help us to better understand the world around us today.

6

CAN THE SUBALTERN SPEAK?

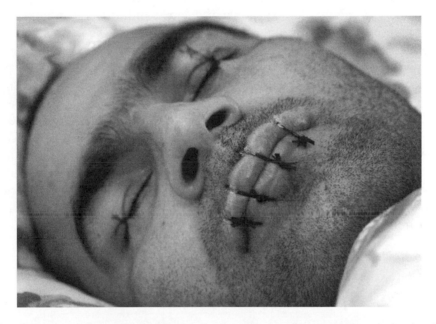

Figure 6.1   'Dutch Iranian immigrant Mehdy Kavousi protests against proposed new asylum laws in Zaandijk, Netherlands, by having his lips and eyes sewn together. On 17 February the Dutch Parliament approved plans by the Christian Democrat Government to expel up to 26,000 failed asylum seekers, despite objections from former Dutch premier Ruud Lubbers, now head of UNHCR. The potential deportees include Somalis, Afghans and Chechens who may be sent back to countries without a functioning government that are still affected by violence.' (*New Internationalist, The Unreported Year 2004*, pp. 8–9)

We have already come into contact with a lot of postcolonial theory in this book, but it has not been labelled as such. In this chapter we will think about it directly, reflecting back throughout to what has been discussed already. Postcolonial theory has made very important interventions in the arts and social sciences in the last 15–20 years, but also has a reputation for being very difficult. It has focused in particular on the ways in which peoples can be known and how this knowledge can be communicated. It is not simply about dry academic matters though. As the opening example demonstrates very visually, marginal people around the world still struggle to have their voices heard, and we shall see how one theorist wishes to incorporate passionate, even violent, emotions and actions into postcolonial practice. We will also examine some of the main aims of postcolonial theory, both in its critique of conventional ways of knowing, and also in its presentation of alternatives.

Postcolonial theory has a number of aims. One of the most important – especially for us as geographers – is its critique of what Edward Said (1978) has named as the two fundamentals of colonial authority, knowledge and power. In other words, the significance of postcolonial theory is a shift in attention from focusing on the economic and political operations of power that allowed western countries to rise to dominance, to understanding the power involved in the continued dominance of western ways of knowing. Because of the networks of power through which western forms of representation of the world circulated, this influenced not only how 'they' were known by 'us', but also how 'they' were persuaded to know themselves. Western ways of knowing – whether this be science, philosophy, literature, or even popular Hollywood movies – have become universalised to the extent that they are often seen as the only way to know. Other forms of understanding and expression are then marginalised and seen as superstition, folklore or mythology.

SPEAKING FOR/AS THE OTHER

The question of the ethics – and even the *possibility* of – speaking for others is one of the major debates in postcolonial studies, including geography. As we have discussed before, Edward Said's classic book *Orientalism* has had an enormous impact on postcolonial studies. He argued that the west's knowledge and representation of the rest of the world were part and parcel of its domination of it; in other words, the west spoke for the other. This idea in turn draws on Foucault's notion of power/knowledge. For Foucault and Said knowledge and power are inseparable; power will be constituted in part through dominant ways of knowledge, which in turn gain their influence through their association with powerful positions within networks.

For knowledge to be powerful it has to be hegemonic, namely it must be accepted to some degree as legitimate, by the ruled as well as the rulers. According to this view everyone has some power, even if only the power to resist, as in the case of the least powerful (the poorest classes, women, tribal groups and other marginalised people, sometimes collectively referred to as 'subaltern'), the power not to believe in what is presented. Mahatma Gandhi made a similar statement when he adopted the strategy of withdrawal of consent in dominant systems of rule (in this case, British colonialism) in non-violent non-co-operation. The others who are represented may sometimes resist interpretations of themselves which demonstrates the important political role within postcolonialism for writing subaltern histories. But who can write such histories? Who can speak for the subaltern?

Gayatri Chakravorty Spivak has written what is often considered the classic essay on the problem of speaking for cultural others, for those whose cultural background is considered profoundly unlike one's own; *Can the subaltern speak?* This is a very difficult article and her conclusion has been interpreted in various ways. Spivak introduces the term 'epistemic violence' which refers to violence done to the ways of knowing and understanding of non-western, indigenous peoples. As we have seen, western ways of knowing have been held up as *the* way of knowing, whether this is in terms of religion, science, philosophy, architecture or governance. Other forms of knowledge have been rendered as less valid, or even downright wrong. Spivak thus offers a critique of the self-assured, scientific approach to studying other cultures that has been characteristic of western scholarship, which makes the western self the subject of history and the nonwestern other its object. She has shown that to be heard – to be listened to and taken seriously – others must also adopt western thought, reasoning, and language. As we have seen throughout this book, this knowledge purported to be an objective view from nowhere. The assumption of the particular experience and perspective of privileged groups should be accepted as universal. But these studies are actually situated in a particular, very powerful, cultural and historical location.

Even when apparently expressing her/his (Spivak talks in terms of 'her') own views, the subaltern is not able to express her true self. Writing about attempts to recover the voices and experiences of subalterns in South Asian historiography, Spivak has argued that the subaltern cannot speak, so imbued must she be with the words, phrases and cadences of western thought in order for her to be heard. In order to be taken seriously – to be seen as having knowledge and not opinion or folklore – the lifeworld of the subaltern is translated into the language of science, development or philosophy, dominated by western concepts and western languages. Thus, as Spivak suggests, the subaltern must always be caught in translation, never truly expressing herself, but always already interpreted. This was clearly expressed during colonialism, when the European language of the colonisers was privileged over local languages. For instance, British colonists believed that their scientific and technical knowledge could be transmitted to Indian students only via classes taught in English. The continued hold of European languages (especially English) through the internet, the media and academic journals demonstrates the relevance of this observation today.

Furthermore, Spivak and others have questioned the degree to which academics and other 'experts' in the west really do want an engagement with people elsewhere, an engagement which would require a decentring of themselves as experts. Bell hooks's (1990) autobiographical approach tells a similar tale to Spivak in her attempt to be heard from the margins of society. For her, the margins are a site of 'radical possibility' (hooks, 1990: 341) which rejects the politics of inside and outside, as 'to be on the margins is to be part of the whole but outside the main body'. It is a knowledge which she believes offers a unique and important perspective that is not distorted by the prejudices of the centre. However, hooks has felt silenced by Western academics who seek the experience, but not the wisdom, of the other. She argues that 'I was made "other" there in that space ... they did not meet me there in that space. They met me at the center' (hooks, 1990: 342). The experiences of the marginalised are used in postcolonial theories, but without opening up the process to their knowledges, theories and explanations. When there is a meeting, it is in the centre – in the (predominantly) Western institutions of power/knowledge (aid agencies, universities, the pages of journals) and in the languages of the west (science, philosophy, social science and so on, expressed in English, French, Spanish ...). So, by approaching the institutions of knowledge, hooks has been forced to the centre; a location both metaphorical in its control of authority and geographical in its physical presence.

She claims to have met not just a reluctance to abandon the mark of authority, but also the desire for material from which explanations can be made. She feels that western researchers want to know about her experiences but not her own explanations:

> No need to hear your voice when I can talk about you better than you can speak about yourself. No need to hear your voice. Only tell me about your pain. I want to know your story. And then I will tell it back to you in a new way. Tell it back to you in such a way that it has become mine, my own. Re-writing you I write myself anew. I am still author, authority. I am still colonizer, the speaking subject and you are now at the center of my talk. (hooks, 1990: 343)

By retelling her experience from a western point of view, hooks's voice is included but only as an example or as data which the Western 'expert' alone can interpret.

Matt Sparke provides the following example of the apparent impossibility of subaltern voices to be heard in modern Canada. His example is of a land claim hearing for the Wet'suwet'en and Gitxsan in 1987. The hearing was transferred from the local court at Smithers, where large numbers of local people could come in support of the claim, to Vancouver, approximately 1200 km away. Few could afford to travel for the case, undermining the collective

Figure 6.2   Can the subaltern speak? (The Law vs. Ayook, taken from M. Sparke (2005) p. 18 *In the Space of Theory*)

action, and further alienating the Wet'suwet'en and Gitxsan from the state proceedings. While they were able to subvert the smooth running of the modern space of court with the performances of their knowledges, ultimately the Wet'suwet'en and Gitxsan voice was not properly heard in the court:

One aspect of the First Nations' subversive courtroom performance was the repeated demonstration of the vitality and importance of their oral histories. Witnesses from both the Gitxsan Wet'suwet'en sung ceremonial songs in court, among them the Wet'suwet'en *kungax* and the many Gitxsan *limx'ooy*, each of which evoked the *adaawk* – a form of historical geographical record – of particular Gitixsan Houses. There was a great deal of controversy in court about having such oral records accepted as legitimate evidence in exemption to the hearsay rule. In a Western juridical field that conventionally accepts only written and cartographic documentation of territory, such oral traditions were cast as illegitimate, and clearly the chief justice showed little respect both in court and in his written *Reasons for Judgement* for what he had heard in this way. When, for example, he was asked whether Antgulilibix (Mary Johnson) could sing a sacred song, McEachern exclaimed:

(Cont'd)

'Could it not be written out and asked if this is the wording? Really, we are on the verge of getting way off track here Mr. Grant. Again I don't want to be sceptical, but to have witnesses singing songs in court is, in my respectful opinion, not the proper way to approach the problem.'

In making his judgement he ultimately reasoned that:

'Except in very few cases, the totality of the evidence raises serious doubts about the reliability of the *adaawk* and *kungax* as evidence of detailed history, or land ownership, use or occupation.'

Nevertheless, the very fact that the songs were sung at all subverted the hushed and sanitized sounds of normal legal procedure [. . .]

In addition to the formal spatiality of the court and its subversion, Monet's cartoon also highlights another more directly cartographic spatial theme. On the one side, he pictures Antgulilibix singing the *limx'ooy*. On the other is the chief justice surrounded by his written records and maps.

(Sparke, 2005: 16–17)

The cartoon brilliantly highlights the inability of the judge, and the Canadian justice system more generally, to understand the First Nations' voice.

But what is the view from somewhere – the situated, authentic minority or the subaltern view? What is the non-hegemonic lived reality and who can authentically represent it? Spivak warns that the recovery of a subaltern voice (especially from history) may involve 'essentialism', a generalised or stereotyped fiction which results in statements like: 'this is "the voice" of Indian tribals/African women/Cairo's poor ... '. Spivak argues that it is very difficult to recover a voice for the subaltern without negating its heterogeneity. She says that the project of finding an effective voice for the subaltern courts the dangers of power/knowledge which the project seeks to avoid above all .

Nevertheless, although Spivak points to complexities and difficulties she does not rule out the possibility of speaking for the other. She raises the possibility of adopting temporary alliances, what she called a '**strategic essentialism**', using a clear image of identity to fight a politics of opposition (which would not be possible if all the aspects of identity were to be incorporated), and so to fight for women's rights, for minority or tribal rights, and so on.

But, in turn, this approach raises the notion of 'authenticity'. How can a privileged Third World male – or even, female – intellectual represent, the poor women of his nation? Can Spivak (a woman) 'speak as', as she puts it, a Third World person, as an Indian woman, as she is so often expected to do in international conferences. When she does try to 'speak as' she worries if she has any right to represent those whose diverse experiences she cannot ever really share. She also worries about 'generalising herself', about reducing herself to an essence, to a set of characterisations, fearing that she will become a token gesture in western academic circles. Will her experience be selected and homogenised in such a way that her audience's ignorance is masked by an *illusion* of knowledge? She says 'you don't hear from all the rest because "we have covered that".' By having her on the programme their guilt is assuaged – that voice has been included – and so some have suggested that postcolonialism is a kind of therapy that is used to overcome anxieties about past injustices and the fact that the effects of these still continue today.

WHAT DOES THIS MEAN FOR HOW WE 'DO' GEOGRAPHY?

Spivak's and others' postcolonial critiques leave many people wondering what they can say. There has been much discussion about whether, if we take postcolonial critiques seriously, western academics can say anything about those from the Third World. The result of this could be the withdrawal of western academics from study outside of the west. This clearly has particularly important implications for the way in which we study geography. However, Spivak herself does not see this as the implication of her critique. She chastises white, privileged, American, male students who feel paralysed by the problems surrounding speaking for an other, when they conclude that, as white males, they should just remain silent. She says to them:

> Why not develop a certain degree of rage against the history that has written such an abject script for you that you are silenced ... make it your task not only to learn what is going on there through language, through specific programs of study, but also at the same time through an historical critique of your own position as the investigating person, then you will see that you have earned the right to criticize, and you will be heard. To refuse to represent a cultural Other is salving your conscience, and allowing you *not* to do any homework.

We can think of the question of whether the subaltern can speak in less theoretical terms. The key is in the infrastructure which produces knowledge about the world: the media, knowledge-producing institutions such as research institutes and universities, and importantly English, French and Spanish as world languages dominate the ways in which we understand the world. These allow the First World to represent the Third World. How can these global structural conditions be opened up to other voices? Perhaps there is the need for 'us' to speak to 'them' in their language? What responsibility if any do First World geographers have for using these channels to represent those who are not yet able to represent themselves? What do you think this might mean for how *you* do fieldwork or study other parts of the world? Consider the last paragraph of bell hooks's paper, '*Marginality as a site of resistance*', which addresses the white academic reader:

> This is an intervention. A message from that space in the margin that is a site of creativity and power, that inclusive space where we recover ourselves, where we move in solidarity to erase the category colonized/colonizer. Marginality as site of resistance. Enter that space. Let us meet there. Enter that space. We greet you as liberators. (hooks, 1990: 343)

POSTCOLONIAL FEMINISM

George W. Bush in large part justified the war in Afghanistan through an appeal for the liberation of Afghan women. Iris Marion Young asserts that Bush's rhetoric here was possible because of feminist campaigns concerning the Taliban, which the Bush administration chose at that moment to exploit:

> I argue that the apparent success of this appeal in justifying the war to many Americans should trouble feminists and should prompt us to examine whether American or Western feminists sometimes adopt the stance of protector in relation to some women of the world whom we construct as more dependent or subordinate ... The women of Afghanistan constituted the ultimate victims, putting the United States in the position of ultimate protector. (Young, 2003: 4, 17)

In a similar vein, Indian novelist and political activist Arundhati Roy (2002) wonders why it is that it is, 'being made out that the whole point of the war was to topple the Taliban regime and liberate Afghan women from their burqas, we are being asked to believe that the U.S. marines are on a feminist mission?' But she points out that there are other places where women are treated very badly (including America's ally, Saudi Arabia, and also parts of South Asia) and so she asks us 'Should they be bombed? Should Delhi, Islamabad and Dhaka be destroyed? Is it possible to bomb bigotry out of India? Can we bomb our way to a feminist paradise?' Roy is pointing to a concern that Third World feminists have had about western liberal agendas, even feminist ones, which is the belief that white men or women are needed to save brown women from brown men (this is often expressed around the politics of the veil – see the box below). They are critical of the western feminist view that all women are united in a sisterhood of exploitation by patriarchy, and that this, rather than divisions based around country, region or religion, must be a woman's prime political concern.

Postcolonial feminists blame western feminists for forgetting the fact that many of the world's women are oppressed by much more than patriarchy; they also face exploitation by global economic systems, race and class. Many Third World women would consider that they have more in common with men of their class and nationality than they have in common with privileged white women. For example, their primary allegiances may be with men in an anti-racist struggle. Many resent the fact that western women have set the agenda of feminism, causing many Third World women to reject the idea of feminism altogether. There is no simple way to add in a consideration of gender with concerns of class or race – one cannot 'add women and stir' or 'add race and stir'. The cross-cutting of oppressions and privileges, of dominations and resistances, is highly complex and very important to the study of postcolonialism.

Many postcolonial feminists thus favour the concept of 'situated knowledge' as a substitute for decontextualised, ungendered, disembodied, so-called 'objective' knowledge. It pays attention to geographic and cultural *specificity* rather than universality. Thus, what it means to be a man or a woman depends upon the context – in other words people learn to become men and women in very diverse material and cultural contexts. Situated feminist knowledge does not pretend to be objective, but claims to be closer to the truth because its interests are not in masking the truth of power relations but instead lie in exposing power structures. Third World critiques of the cultural imperialist attitudes of western feminists have opened the way for feminism to concern itself with simultaneous oppressions as well as with racism, neo-colonialism, the forced migration of ex-colonial populations, and global economic structures. Spivak and other Third World feminists challenge the coherence of geographical boundaries to argue that there is a Third World within the First World that knows the complexity of oppression better than any privileged women would understand. As a result, white women are beginning to study racism and the social construction of whiteness as an approach to feminist understanding. There is a growing recognition that 'races' and genders are defined in relation to each other, a fact that is often much more obvious to subordinate groups than those who benefit from the privileges of dominance. Ruth Frankenberg has highlighted the fact that, for most white people, 'race' is something that others are marked by. Even for those critical of racism, 'race' is the mark of discrimination, of repression and of poverty. What Frankenberg does is to demonstrate that whiteness is also a racial marker which signifies a hidden privilege that reinforces it as a marker that white people have; 'a privilege enjoyed but not acknowledged, a reality lived in but unknown' (Thomas, quoted in Frankenberg, 1993: 51).

Such postcolonial critiques suggest that it is difficult to operate a conventional politics of identity and opposition – the complexity of identity and allegiance means that only strategic alliances make any sense. Another postcolonial response is to abandon the stable politics of identity and opposition, for one based instead upon uncertainty and ambivalence. Such a politics is talked of in terms of 'hybridity'.

THE CULTURAL POLITICS OF THE VEIL

Of all of the symbols of Orientalism, the veil has been that which has most greatly concerned, intrigued, fascinated and troubled western commentators. European explorers noted the practice of veiling, seeing it as both a backward covering of women as well as being intrigued by what the veil might conceal. Some commentators have written about a western desire of freeing women from the prison of their veil (a desire that Spivak (1994) has talked about as 'white men wanting to save brown women from brown men'). We can see the complexity of the veil by considering recent discussions of it.

(Cont'd)

Figure 6.3   *Indigène Du Caire*

The issue of the meaning of veils was discussed in the media and parliament in the UK in the autumn of 2006, as a result of the decision made by Kirklees Council in Dewsbury to suspend classroom assistant Aishah Azmi for refusing to remove her full face veil at school. The discussion expanded to consider the meaning of the veil in modern, multicultural Britain. Some, including members of the Labour government, felt that the veil was a barrier to the inclusion and acceptance of British Muslims into wider society. Others felt that a multicultural society must be willing to recognise difference and accept the beliefs of minority groups.

In France, the decision made a few years previously by the government to ban the wearing of headscarves by Muslim girls at public school – along with large crosses and other 'overt' religious symbols – has apparently re-inscribed the body of the citizen as a neutral space, 'maintaining a distance between all spiritual or community affiliations in the public arena, thus making all equal and the same in civic life' (Lévy, 2004:5.2). It is interesting that despite the list of religious artefacts mentioned, it is the veil that has aroused so much interest and debate. This is the image of the exotic, seductive Orient that has so long intrigued western imaginations, through the desire to reveal what lies behind at the same time as the desire to free women from what is seen as religious indoctrination or the exercise of patriarchal power. Once again, there is a battle over the territory of women's bodies – the secular versus the religious. But where are the women themselves in this debate? When asked many claim it as part of their own identity, as a symbol of a global sisterhood of Muslims rather than a patriarchal sign; something that, for some, has become all the more important since 9/11. As one British Muslim woman explained, 'In many ways I saw the *hijab* as an act of solidarity with Muslim women all around the world (Aziz, 2004).

Many outside commentators have noted that there is an increasing tendency for women in the Middle East to 'return to the veil'. Western commentators have assumed that this has been due to reactionary forces seeing women as symbolic of the nation, and thus as repositories for the traditional image. However, this is to simplify the meaning of the veil.

The Qur'an does not explicitly state that women should be veiled, although it does require them to cover their breasts and ornaments. So the veil is not a part of the religious laws of Islam but instead is based around cultural values concerning the modesty of women. However, the reasons that women give for wearing the veil are multiple and can be about choice as well as a feeling that it is a requirement; it can be seen as a symbol of patriarchal power or as a symbol of resistance to westernisation; it can be a symbol of religious adherence or something which allows women to blend in. Also, we must not forget that the veil is not only a symbol of Islam, but has been used by other religions too (think of Christian nuns).

Many independent and capable women are veiled. The western view of modernised woman versus passive veiled instrument of tradition needs to be challenged. It seems that there are many 'gender performances' in Islamic societies which may end up with virtually the same image – indeed there are those for whom reveiling represents a resistance to the values of the west. However, in her research in Egypt, Sherifa Zuhur (1992) discovered that others use the image of the veiled woman to negotiate the public spaces of work to present a non-threatening image to male workers while transgressing the public-private divide – a strategic use of clothing to allow women greater freedom while under the guise of an image of the 'good woman'. For other women, more straightforwardly, it seems it has become a fashion, with the colour and material of the headscarf being chosen with the same amount of attention given as with the remainder of the outfit.

But, of course, wearing the veil does not necessarily mean the same thing everywhere. While the *hijab* can be a symbol of pride for being part of a Muslim sisterhood in places where there is the choice to do so, where this represents an act of defiance or a clear statement, in

(Cont'd)

Figure 6.4 *Hijab* as fashion

other places women may not have the choice whether or not to make this statement. In such places a very different identity politics emerges, where the territory of women's bodies is subject to different forces. One possible clear outcome of the French example is that those who want their daughters to continue to remain veiled (and who can afford to) will send them to special Muslim schools, hence reinforcing divisions between communities in the future, and more clearly still inscribing territories of belonging and exclusion.

Discussing her return from Iran, Iranian-American Azadeh Moaveni, hints at the complex meaning of the veil:

Today, in a quiet room in a country not far from Iran in space, I am finally unpacking the boxes from those two years in Tehran. As I sort through the clothes, peeling veil from veil, it is like tracing the rings of a tree trunk to tell its evolution. The outer layers are a wash of color, dashing tones of turquoise and frothy pink, in delicate chiffons and translucent silks. They are colors that are found in life — the color of pomegranates and pistachio, the sky and bright spring leaves — in fabrics that breathe. Underneath, as I dig down, there are dark, matte veils, long, formless robes in funeral tones of slate and black. That is what we wore, back in 1998. Along the way, the laws never changed. Parliament never officially pardoned color, sanctioned the exposure of toes and waistlines. Young women did it themselves, en masse, a slow, deliberate, widespread act of defiance. A jihad, in the classical sense of the word: a struggle. (Azadeh Moaveni (2005), *Liptick Jihad: A Memoir of Growing up Iranian in America and American in Iran*, p. ix)

HYBRIDITY

Hybridity is the name of this displacement of value … that causes the dominant discourse to split along the axis of its power to be representative, authoritative. (*Bhabha, 1994: 113*)

Rather than directly opposing colonial or dominant discourses, some postcolonial theorists such as Homi Bhabha redefine postcolonial positively as uncertainty, ambivalence, hybridity, a third space and the space of multiple cultural borders. Rather than basing truth on authenticity, it valorises impurity and mixing.

This presents a profound challenge to the geography of Orientalism. Whereas, as we have seen, modern western thought is structured around a series of binaries which suggest that a person is one thing or the other (black or white, Oriental or Occidental, deviant or normal, and so on), postcolonial hybridity looks to the grey areas, arguing that there is no neat inside/outside division. The colonised are like colonisers in some ways (culture, education, language, and so on) but are also different (colour, origin). The very existence of this mixing – the mark of the self in the other – challenges the existence of binaries. Thus, the binary logic of Orientalism *cannot work* with the existence of such ambiguity: the very existence of the hybrid highlights the binaries as fantasy. Bhabha argues that the discovery of hybridity offers the emergence of resistance to colonialism, which ultimately in turn heralds the end of colonialism because it has not been possible to keep Occidentalism pure and separate from Orientalism.

Bhabha argues this by moving from the texts of colonialism (maps, manuals, official records and literature) to consider the actual practices of colonialism which, unlike these texts, were messy, incomplete and sometimes incompetent. The daily reality of colonial practice ensured that the purity of colonial ideas could not exist in actuality. And we might add as geographers that the differentiated landscapes upon which colonialism was practised would also ensure that reality would not be the same as theory – for instance, particular physical landscapes would offer different challenges to the projects of building colonial settlements or plantations, different cultural practices would respond differently to colonial rules and regulations and so on.

For Bhabha then, the very enactment of colonialism heralded its end. Rather than seeing impurity/mixing as bad (as colonial administrators did) or as a loss (as in the salvage paradigm), Bhabha celebrates the hybrid as the figure of post-colonialism. For Bhabha, the hybrid offers a resistant politics that does not simply redraw boundaries (as Said could be accused of doing by accepting the division between the Occident and Orient) but subverts them altogether.

For many theorists, hybridity has come to represent the icon of postcolonialism. As something which celebrates ambivalence and impurity, it offers a profound challenge to the colonial logic which attempted to catalogue and know the world. This has lead some to fear that hybridity was being uncritically celebrated. Katharyne Mitchell (1997) calls this the 'hype of hybridity'. She argues that the privileging of hybridity celebrates everything that is impure because it sees essences and authenticity as bad as they lead to boundaries, binaries, hierarchies and exclusions. However, she also argues that the powerful can hybridise as well. She uses the example of international capitalism, which very effectively exploits mixtures and new forms (of identity, culture, food, music, fashion …) in its constant bid for new products and thus more profit. Furthermore, Mitchell suggests that not all bids for purity or authenticity are bad. As we have seen already, essentialisms can be politically useful – ambivalence can seem like a weaker form of political action than outright opposition. Indeed some would argue that the politics of hybridity (such as mimicry) are somewhat trivial when compared to political marches, demonstrations, and the other embodied consequences of outright opposition which was seen over and over again in opposition to colonial administrations.

Mitchell's conclusion is that there is no *inherent* value to either hybridity *or* essential identity, and that we must judge each case based on the issues involved (and our own interpretation of the events). Others, however, have argued more powerfully about the negative effects of colonial hybridity, seeing this as a condition of trauma rather than something to be celebrated.

FANON AND THE VIOLENCE OF POSTCOLONIALISM

Some forms of postcolonialism have suggested that a more direct form of intervention is required for change, and one prominent theorist, Frantz Fanon, has even called for the necessity of violence. Violence is a very difficult issue to discuss as, by its very nature, it is destructive and divisive. Some postcolonial theorists and political figures insist that there is never a place for violence, while others have suggested that there are occasions when people are otherwise so powerless that this is the only door left open to them, their acts of violence being the only 'voice' they have that will be heard by those in control (think about the current debates around suicide bombers in the Middle East).

Fanon had a very different interpretation of the colonial world than Bhabha, although he too recognised the blurring of boundaries between coloniser and colonised. For Fanon, though, hybridity was a very destructive thing which had to be resisted head on.

In his 1963 book, *The Wretched of the Earth*, Fanon argued that the 'colonial world is a Manichean world', a world of black and white alone, wherein colonial knowledge sought to completely separate black from white. Within this binary, Fanon argued that two elements of colonialism were in tension:

1   Through colonial rule and education, the colonised were constantly told of the superiority of colonisers' values and that these should be aspired to and copied.
2   However, at the same time there was the existence of what Fanon called 'the fact of blackness', in other words, the manner in which a colonised person can most immediately be identified is by the colour (or other features) of their skin. This 'fact of blackness' was a marker of inferiority which was inescapable.

For Fanon then, colonial authority worked by inviting black subjects to mimic white culture, an argument he developed in his book *Black Skin, White Masks* (1967). But for the colonised, the first point – the copying of colonisers' values – can never be totally successful, and it is this which maintains the superiority of the colonisers. Loomba explains this argument as follows: 'Indians can mimic but never exactly reproduce English values … their recognition of the perpetual gap between themselves and the "real thing" will ensure their subjection' (Loomba, 1998: 173).

The colonised will then be trapped between the 'fact' of their racial difference, and their aspirations to the colonisers' culture which has been held up as the model for civilisation and development. They will always remain mixtures or hybrids. However, rather than celebrating this ambivalent position as other postcolonial thinkers like Bhabha do, Fanon sees it as a traumatic condition. We can best realise how Fanon came to this conclusion by looking at his own life.

Fanon was born in Martinique and was initially the model colonised subject. He volunteered for military service in the Second World War and afterwards returned to France to train as a doctor. However, when he subsequently went to Algeria to work for the French, he resigned because of the brutality of the colonial regime. Through his work as a medic he was exposed to the psychological impacts of colonialism that had resulted in a range of mental health problems in his patients. Because of this, he joined the Algerian resistance and wrote his treatises against colonialism.

Fanon argued that psychic trauma occurred when the colonised subject realised that he (and it was always 'he', something we will return to shortly) would never be able to attain the whiteness he had been taught to desire, or shed the blackness he had learnt to devalue. His conclusions were stark: 'At the risk of arousing the resentment of my coloured brothers', he argued in

Black Skin, White Masks, 'I will say that the black man is not a man'. This was due to the impacts of the colonial system:

> Because it is a systematic negation of the other person and a furious determination to deny the other person all attributes of humanity, colonialism forces the people it dominates to ask themselves the question constantly: 'In reality, who am I?' (Fanon, 1963: 250)

BLACK SKIN, WHITE MASKS

Figure 6.5 Two interpretations of *Black Skin, White Masks*. What is the nature of the masks in each?

Colonialism had destroyed the black subject. For the white subject, the black other is everything that lies outside of the self (identity is defined by projecting negative characteristics onto the other, as explained by Said in *Orientalism*). For the black subject, however, the white other serves to define everything that is desirable: the white subject is not just other, but also master (because of unequal power relations). Blackness confirms the white self, but whiteness empties the black subject.

The black person attempts to cope by adopting white masks that will somehow make the fact of his blackness vanish. There are various colloquial expressions for this, such as the Australian aboriginal term 'coconut'. Black skin/white masks reflects the miserable schizophrenia of the colonised's identity.

Fanon produced an analysis of the psychological problems facing black men and women in a white world. The condition he identified was self-division or alienation from the self. He presented a number of case notes recording 'colonial war and mental disorders' (1963). One diagnosed a man's sexual impotence as being the trauma caused by the rape of his wife by colonising soldiers (see the box below). As a response to these traumas, Fanon argued that the colonised need 'nothing short of the liberation of the man of colour from himself'.

FANON AND THE TRAUMA OF COLONIALISM

In the chapter 'Colonial war and mental disorders', in *The Wretched of the Earth*, Fanon provides information on a number of cases he examined between 1954 and 1959 in Algerian hospitals. Most involve Algerians, but Fanon also considers the impact of the war on French individuals as well.

Consider the case below which describes man B—'s trauma, and his ability to narrate his story after receiving therapy. Think about the ways in which the war had become internalised by this individual through his own actions, and actions done to his wife, but also think critically about how Fanon is telling the story. Why is the story about the man? What role does the wife play in this story? What about the trauma she suffered?

Case No. 1: Impotence in an Algerian following the rape of his wife

B— is a man twenty-six years old. He came to see us on the advice of the Health Service of the FLN [*Front de Libération Nationale*] for treatment of insomnia and persistent headaches. A former taxi-driver, he had worked in the nationalist parties since he was eighteen. Since 1955 he had been a member of a branch of the FLN. He had several times used his taxi for the transport of political pamphlets and also political personnel. [...]

One day however, in the middle of the European part of the town, after fairly considerable fighting a very large number of arrests forced him to abandon his taxi, and the commando unit broke up and scattered. B—, who managed to escape through the enemy lines, took refuge at a friend's house. Some days later, without having been able to get back to his home, on the orders of his superiors he joined the nearest band of Maquis [resisters].

For several months he was without news of his wife and his little girl of a year and eight months. On the other hand he learned that the police spent several weeks on end searching the town. After two years spent in the Maquis he received a message from his wife in which she asked him to forget her, for she had been dishonored and he ought not to think of taking up their life together again. He was extremely anxious and asked his commander's leave to go home secretly. This was refused him, but on the other hand measures were taken for a member of the FLN to make contact with B—'s wife and parents.

Two weeks later a detailed report reached the commander of B—'s unit.

His abandoned taxi had been discovered with two machine-gun magazines in it. Immediately afterward French soldiers accompanied by policemen went to his house. Finding he was absent, they took his wife away and kept her for over a week.

(Cont'd)

She was questioned about the company her husband kept and beaten fairly brutally for two days. But the third day a French soldier (she was not able to say whether he was an officer) made the others leave the room and then raped her. Some time later a second soldier, this time with others present, raped her, saying to her, 'If ever you see your filthy husband again don't forget to tell him what we did to you.' [. . .] When she told her story to her mother, the latter persuaded her to tell B— everything. Thus as soon as contact was re-established with her husband, she confessed her dishonor to him. Once the first shock had passed, and since moreover every minute of his working time was filled with activity, B— was able to overcome his feelings. [. . .]

In 1958 he was entrusted with a mission abroad. When it was time to rejoin his unit, certain fits of absence of mind and sleeplessness made his comrades and superiors anxious about him. His departure was postponed and it was decided he should have a medical examination. This was when we saw him. He seemed at once easy to get to know; a mobile face: perhaps a bit too mobile. Smiles slightly exaggerated; surface well-being: 'I'm really very well, very well indeed. I'm feeling better now. Give me a tonic or two, a few vitamins, and I'll build myself up a bit.' A basic anxiety came up to break the surface. He was at once sent to the hospital. [. . .] at the end of several days we were able to reconstruct his story.

During his stay abroad, he tried to carry through a sexual affair which was unsuccessful. Thinking that this was due to fatigue, a normal result of forced marches and periods of undernourishment, he again tried two weeks later. Fresh failure. Talked about it to a friend who advised him to try vitamin B-12. Took this in form of pills; another attempt, another failure. [. . .]

[H]e spoke to us for the first time about his wife, laughing and saying to us: 'She's tasted the French.' It was at that moment that we reconstructed the whole story. The weaving of events to form a pattern was made explicit. He told us that before every sexual attempt, he thought of his wife. [. . .]

'In the Maquis, when I heard that she'd been raped by the French, I first of all felt angry with the swine. Then I said, "Oh, well, there's not much harm done; she wasn't killed. She can start her life over again." And then a few weeks later I came to realise that they'd raped her *because they were looking for me*. In fact, it was to punish her for keeping silence that she'd been violated. She could have very well told them at least the name of one of the chaps in the movement, and from that they could have searched out the whole network, destroyed it, and maybe even arrested me. That wasn't a simple rape, for want of something better to do, or for sadistic reasons like I've had occasion to see in the villages; it was the rape of an obstinate woman, who was ready to put up with everything rather than sell her husband. And the husband in question, *it was me*. This women had saved my life and had protected the organization. It was because of me that she had been dishonored. And yet she didn't say to me: "Look at all I've had to bear for you." On the contrary, she said: "Forget about me; begin your life over again, for I have been dishonored." [. . .]

'So I decided to take her back; but I didn't know at all how I'd behave when I saw her. And often, when I was looking at the photo of my daughter, I used to think that she too was dishonored, like as if everything that had to do with my wife was rotten. If they'd tortured her or knocked out all her teeth or broken an arm I wouldn't have minded. But that thing – how can you forget a thing like that? And why did she have to tell me about it all?'

Fanon, *The Wretched of the Earth*, (1963) pp. 254–58

Fanon reversed the ideas of the colonisers. Rather than seeing colonialism as the *result* of the failings of the other (namely, European colonialism was needed as the others were not able to rule themselves), Fanon insisted we now saw it as the *cause* of the mental defects of the other. For Fanon, it was colonialism that was regarded as psychopathological and violent, not the colonised. Thus, the negative character traits that were associated with the colonised (violence, hysteria, laziness …) were interpreted by Fanon as conditions brought on by colonialism.

In addition to the psychological examination of the conditions of colonialism, Fanon thought of what should be done. His vision for action was laid out in *The Wretched of the Earth: The Handbook for the Black Revolution that is Changing the Shape of the World*. In this book, Fanon discusses the inevitability of violence in decolonisation.

Why violence?

Read Fanon: you will learn how, in the period of their helplessness, their mad impulse to murder is the expression of the natives' collective unconscious. (John-Paul Sartre (from the introduction to *The Wretched of the Earth*, 1963: 18)

Europeans saw anti-colonial resistance as incredibly violent, illustrating the necessity for a European, civilising influence and the lack of ability of the indigenous population to rule themselves. But for Fanon it was colonialism itself that was the real source of violence; firstly in the initial act of conquest, but then in the day-to-day practices of rule which required a power over the body and mind of the native. For Fanon, this was a totalising violence, which, as Sartre explained, 'does not only have for its aim the keeping of these enslaved men at arm's length; it seeks to dehumanize them' (1963: 15). Fanon also argued that this dehumanisation meant that colonised societies would become violent. The alienation created by colonialism would first generate violence among the colonised. Their frustration meant this initial violence would erupt between those who were colonised. However, Fanon saw that it was important that this violence was turned against the colonisers, arguing that this would fulfil a number of purposes. Firstly, it would prise loose the grip of the colonists, and as the colonisers would not want to give up power, decolonisation would always be a violent phenomenon. Secondly, the actions of decolonisation would bring together the colonised in a common purpose. Finally, and most controversially, violence would be cleansing and restorative for those individuals subject to colonial rule, purging their feelings of impotence and inferiority and restoring their self-respect.

At the level of individuals, violence is a cleansing force. It frees the native from his inferiority complex and from his despair and inaction; it makes him fearless and restores his self-respect … When the people have taken violent part in the national liberation they will allow no one to set themselves up as 'liberators' … Yesterday they were completely irresponsible; today they mean to understand everything and make all decisions. (Fanon, 1963: 94)

STRATEGIES OF RESISTANCE

Fanon offered strategies of resistance that played with tradition, but more importantly with European perceptions of it. A particularly interesting example is 'Algeria Unveiled'. The wearing of veils by Algerian women was one strategy developed in opposition to colonialists who were determined to unveil Muslim women in the name of progress, but it was also part of a campaign to weaken indigenous culture and destroy anti-colonial resistance. For many women, the veil was worn because the French wanted them to remove it. But it also gave women a strategy in the resistance movement: unveiled women were unremarkable (because they were western looking) so they could penetrate the European quarter. At the same time, veiled women could transport weapons under their voluminous clothing. Both strategies were used in the anti-colonial movement in Algeria, with women changing their appearance to move freely around the city.

This is illustrated in one of the most powerful scenes in *The Battle of Algiers*, where we see veiled women removing their veils and changing their appearance to look western and get through one checkpoint, then adopting the veil again to get through another, in both cases to place bombs in public places. The operation of cultural practices is both the occasion for resistance and the means of resistance, and also shows the ways in which colonial oppression produces some of the weapons that are used against it.

Figure 6.6   Two images of Algerian women from the film

Fanon argued that the traumas he uncovered did not simply disappear with decolonisation unless there was this violent overthrow to allow the colonised to regain their dignity and sense of self. A peaceful handover of power ran the risk of simply replacing the European colonisers with westernised local elites rather than with the more profound changes which he felt would emerge from armed struggle. Nothing would alter. The people at the top would be different, but life – including the sense of psychological inferiority that Fanon discussed – for the majority of people would continue as it had done. This would therefore mean that even after physical decolonisation (the removal of Europeans) the minds of the people would remain colonised. Thus, Fanon insisted that this perpetuated a sense of ambiguity that would continue due to the power of neo-colonialism, of Western-dominated international capital, that would not allow newly independent countries to develop in peace.

Fanon's argument here is a powerful one. His books are written with the passion and authority of someone who was involved in the Algerian resistance. The arguments he makes still clearly resonate today in terms of debates over independence movements and the appropriate actions that should be taken. Reading Fanon gives eerie resonances with recent news reports about suicide bombers and violent revolts in Palestine, Chechnya and Iraq, amongst others. However, a number of important issues still need to be evaluated in terms of Fanon's arguments. Most obvious is the role of violence. I am very uncomfortable with any arguments which advocate violence. And yet, I have never lived in a colonised society, I have never faced the traumas which Fanon describes, I have never experienced the sense of complete hopelessness that his case notes portray, and so, perhaps it is too easy for me to condemn violence outright.

Fanon's vision of change is also not without its flaws. For instance, as we have seen there is a gendering of trauma in his writing. His books are written for male resisters. If we look at a passage from *The Wretched of the Earth* this is abundantly clear:

> The look that the native turns on the settler's town is a look of lust, a look of envy; it expresses his dreams of possession – all manner of possession: to sit at the settler's table, to sleep in the settler's bed, with his wife if possible. (Fanon, 1963: 39)

Think also of the example of the man who was traumatised through the rape of his wife. It was not the violence to the woman which was the focus of Fanon's concern but the psychological trauma that this had on her husband. It was as a result of his inability to protect his wife that he became impotent, Fanon insisted. Thus, as we see from the above, the solution is a violent revolution which allows the colonised to turn the tables. This vision of women as objects of sexual violence is thus deeply problematic.

Further reading

On introducing postcolonial theory

Childs, P. and Williams, P. (1997) *An Introduction to Post-colonial Theory*. London: Prentice Hall.
Loomba, A. (1998) *Colonialism/Postcolonialism*. London: Routledge.

On speaking for/as the other

hooks, b. (1990) 'Marginality as a site of resistance', in R. Ferguson et al. (eds), *Out There: Marginalization and Contemporary Cultures*. Cambridge, MA: MIT. pp. 341–43.
Spivak, G.C. (1994) 'Can the subaltern speak?' in P. Williams, and L. Chrisman (eds), *Colonial Discourse and Post-colonial Theory*. New York: Columbia. pp. 66–111.

On how we 'do geography'

Huggins, J., Huggins, R. and Jacobs, J. (1995) 'Kooramindanjie: place and the postcolonial', *History Workshop Journal*, 39: 165–81.
An example of work which seeks to bring in different voices in this case of Australia is this collaborative piece about a visit to Carnarvon Gorge National Park by geographer Jane Jacobs and Aboriginal historian Jackie Huggins, her mother and her young son.
Radcliffe, S. (1994) '(Representing) post-colonial women: authority, difference and feminisms', *Area*, 26: 25–32.

On postcolonial feminism

Mohanty, C., Russo, A. and Torres, L. (eds) (1991) *Third World Women and the Politics of Feminism*. Bloomington, IN: University of Indiana Press.

On hybridity

Bhabha, H. (1994) *The Location of Culture*. London: Routledge.
Mitchell, K. (1997) 'Different diasporas and the hype of hybridity', *Environment and Planning D: Society and Space,* 15: 533–53.

On Fanon

Fanon, F. (1963) *The Wretched of the Earth*. Harmondsworth: Penguin. (Especially 'Concerning violence'.)
Sartre, J-P. (1963) 'Preface' in F. Fanon, *The Wretched of the Earth*. Harmondsworth: Penguin.

Finally, watch *The Battle of Algiers* and think about the issues of violence raised by Fanon. How do these resonate with the violence we see in contemporary wars and resistance movements (for example, the war in Iraq or suicide bombers in Palestine)?

7
POSTCOLONIAL CULTURE

Like a failing bus labouring through the sky, the Gulf Air plane seemed barely to be managing, though most of the passengers felt immediately comfortable with this lack of oomph. Oh yes, they were going home, knees cramped, ceiling level at their heads, sweat-gluey, fate-resigned, but happy.

The first stop was Heathrow and they crawled out at the far end that hadn't been renovated for the new days of globalization but lingered back in the old age of colonization.

All the third-world flights docked here, families waiting days for their connections, squatting on the floor in big bacterial clumps, and it was a long trek to where the European-North American travelers came and went, making those brisk no-nonsense flights with extra leg-room and private TV, whizzing over for a single meeting in such a manner that it was truly hard to imagine they were shitting-peeing, bleeding-weeping humans at all. Silk and cashmere, bleached teeth, Prozac, laptops, and a sandwich for their lunch named The Milano.

Kiran Desai, *The Inheritance of Loss (2006)*, p. 285

In this chapter we will consider how some of the ideas from postcolonial theory are paralleled in cultural products such as novels, music and films. We will seek to include the voices of artists from outside the west (or those whose identities stretch between the west and the rest) to see how they have explored, and have attempted to express, the postcolonial ideas of voice, gender relations and hybridity.

We have already discussed the western consumption of postcolonialism and postcolonial products and questioned suggestions that, despite imperial power relations, it is not simply a one-way process. Non-western others have agency to represent themselves too. The question, as raised in the previous chapter, is how these voices are raised, heard and allowed to challenge dominant knowledge and cultural systems. There are forms of music, literature, art and film which present alternative sites of representation to western culture. These have been caught up in global processes of distribution and consumption so that they reach a wider audience, but they also offer different reactions to globalisation and cultural imperialism. Some attempt to preserve cultural authenticity, while others choose a celebration of hybridity.

All the products we will discuss here have been pulled from the media and as a result are necessarily selective. Spivak would caution us about the extent to which we can regard these as the voices of the other.

HYBRIDITY

As discussed in the previous chapter, many postcolonial theorists challenge the binary identities of self and other and this is also expressed within the hybrid forms of postcolonial culture. This notion of hybridity is not just the bringing together of two cultures, but is also the creation of something new out of difference, offering the possibility of a third way or a 'Third Space' that is not of the centre or of the periphery, or inside or out, or the developed world or the developing world, but is a space that, lying as it does in both, negates the possibility of such dualisms. In the previous chapter we saw how this worked out in terms of postcolonial theory, but there is also a multitude of examples in music (such as the work of Natacha Atlas, Claude Challe, Youssou N'Dour and Ali Farka Touré), Bollywood films, and novels (we could think of authors like Ahdaf Soueif, Hanif Kureshi and Kiran Desai).

Someone who has worked with the concept of the hybrid is Salman Rushdie, an Indo-Englishman whose writing reflects this hybrid national identity. He argues that hybrid writing has now become a necessity because English is 'no longer an English language, now grows from many roots; and those whom it once colonised are carving out large territories with the language for themselves' (Rushdie, 1982: 10)

This is represented in his works in many ways. including the use of South Asian words in the text, references to both western and Indian stories and legends, and the location of his novels in real and imaginary places. In his most (in)famous novel, *The Satanic Verses,* hybridity is centred in the text in the figures of the two protagonists who fall out of a plane that has exploded due to a terrorist bomb at the opening of the story and land on a beach on the south coast of England. Their dramatic arrival clearly symbolises recent post-colonial 'invasions' of England and Englishness as a unitary identity. Throughout the novel he insists that the legacy of colonialism negates the possibility of *an* Englishness. As we have seen from earlier chapters, colonisers superimposed European knowledge and value systems upon indigenous peoples across the globe. How then could either society remain culturally 'pure' after this colonial contact?

The *Satanic Verses* explores the effects of people being produced by more than one culture: people as the products of globalisation, an ever integrating world system of politics, economics, culture and identity. Rusdie's texts are rich with the ambivalences, contradictions and sometimes bizarre juxtapositions of modern life: references to western TV images and Indian literature, traditional songs and movie themes sit side by side in his narration of contemporary globalised societies. He also explores the different consequences of the hybridising processes of contemporary society for his protagonists. One character wants to be a real, 'authentic' Indian (Gibreel Farishta), while another wishes to be accepted in England, as someone celebrating hybridity, a real brown-Englishman (Saladin Chamcha):

> … Chamcha is a creature of *selected* discontinuities, a *willing* re-invention, his *preferred* revolt against history being what makes him … 'false'? … While Gibreel, to follow the logic of our established terminology, is to be considered 'good' by virtue of *wishing to remain*, for all his vicissitudes, at bottom an untranslated man … such distinctions, resting as they must on an idea of the self as being (ideally) homogeneous, non-hybrid, 'pure,' – an utterly fantastic notion! – cannot, must not suffice. (*Rushdie, The Satanic Verses* (1988): 427)

A central question that the novel raises is who is the 'real' person? Who is truthfully being himself and who is the invention? The novel suggests that neither alone are 'true', that both are equally real and equally fabrications. Indeed Rushdie suggests that they are two sides of the same coin and cannot be separated, suggesting that the duality 'A AND B' is the postcolonial condition, rather than a colonial binary 'A NOT B', just as the postcolonial theorists we discussed in the previous chapter had decided.

Rushdie presents this book as a literal equivalent of the migrant:

> If *The Satanic Verses* is anything, it is the migrants' view of the world. It is written from the very experience of uprooting, disjuncture and metamorphosis … that is the migrant condition, and from which, I believe, can be derived a metaphor for all humanity. (Rushdie, quoted in Bhabha, 1990: 16)

His hybrid geography is conveyed in his novel in a number of ways:

1 *In the story, which is located in England and South Asia.* However, Rushdie stresses the similarities between these distant places rather than the coherence and unity of bounded space. He challenges the perceived isolation of places and the spatial divisibility of character:

> How far did they fly? Five and a half thousand as the crow. Or: from Indianness to Englishness, an immeasurable distance. Or, not very far at all, because they rose from one great city, fell to another. The distance between cities is always small; a villager, traveling a hundred miles to town, traverses emptier, darker, more terrifying space. (*Rushdie, The Satanic Verses* (1988): 41)

2 *In references to specific instances of both of the cultures that make up the story.* These are significant moments in British history, such as the Norman Conquest and contemporary characteristics of British culture, but also Indian history,

mythology and movie stars. Depending on your background, sometimes you will recognise the reference, sometimes you will realise you do not, and at other times you will not even be aware there is a reference. Rushdie is therefore creating a sense of hybridity for the reader: sometimes you become part of the community, sometimes you know you've been left out – like the migrant both within and without.

3   *In the language which includes Indian words, terms and names from the Qu'ran.* For example, many character names have meanings in Arabic but simply appear as exotic signifiers to a reader who does not know that language: Jahilia, Rushdie's name for the Holy City in *The Satanic Verses*, means 'ignorance' in Arabic and refers to the name for the period before Islam; Chamcha, the name given to one of the protagonists in the same novel, means 'ass-kisser'. For those who know these references, a deeper sense of belonging is established; for those who do not, there is a sense of otherness, exclusion and difference evoked in the book.

4   *Points 2 and 3 mean that there is a geography for the readership of Rushdie's work.* Different readers achieve different interpretations and readings of the text depending on their cultural backgrounds, each reader being included in some references and excluded from the meaning of others.

5   *In the non-linear narrative which moves forwards and backwards through hundreds of years*: the period when the Qur'an was dictated to Mohamed is mixed in with contemporary events, refusing to accept the secular progressiveness of history. Fantastic characters combine myth and reality. The two main characters arrive in modern day Britain to find that they have metamorphosed into chimerical beasts and wander around London with people responding to them as such. This exposes the power of stereotypical images being projected onto their bodies by a society still dependent upon Orientalist views of the world. The protagonists are hybrids of British and Indian culture but in contemporary racist Britain, they are seen to be foreign and Other: Orientalist stereotypes of Indian people as fantastic but frightful characters (see the box below). For Rushdie as a postcolonial migrant, the real and metaphorical do not exist in separate worlds: the symbolic and literal are in part constitutive of each other.

THE MIGRANT EXPERIENCE IN *THE SATANIC VERSES* (PP.167–8)

One of the protagonists, Saladin Chamcha, finds himself transformed, from his forehead have sprung horns, he has developed awful, sulphurous bad-breath, his legs have become powerful and incredibly hairy, and his feet have been replaced with hooves. After a run-in with local police, he awakens in an institution populated by similarly magical beings.

> … Saladin was awakened by a hiss out of an Indian bazaar.
> 'Ssst. You, Beelzebub. Wake up.'
> Standing in front of him was a figure so impossible that Chamcha wanted to bury his head under the sheets; …

It had an entirely human body, but its head was that of a ferocious tiger, with three rows of teeth. ...

Just then a voice from one of the other beds ... wailed loudly: 'Oh, if ever a body suffered!' and the man-tiger, or manticore, as it called itself, gave an exasperated growl. 'That Moaner Lisa,' it exclaimed. 'All they did to him was make him blind.'

'Who did what?' Chamcha was confused.

'The point is,' the manticore continued, 'are you going to put up with it?'

Saladin was still puzzled. The other seemed to be suggesting that these mutations were the responsibility of – of whom? How could they be? – 'I don't see,' he ventured. 'who can be blamed ...'

The manticore ground its three rows of teeth in evident frustration. 'There's a woman over that way,' it said, 'who is now mostly water-buffalo. There are businessmen from Nigeria who have grown sturdy tails. There is a group of holidaymakers from Senegal who were doing no more than changing planes when they were turned into slippery snakes. I myself am in the rag trade; for some years now I have been a highly paid male model, based in Bombay, wearing a wide range of suitings and shirtings also. But who will employ me now? he burst into sudden and unexpected tears. ...

'But how do they do it?' Chamcha wanted to know.

'They describe us,' the other whispered solemnly. 'That's all. They have the power of description, and we succumb to the pictures they construct.'

Think about the parallels between this paragraph and the ideas of colonial power/knowledge we read about earlier, and compare this breakdown of categories with Foucault's description of the 'certain Chinese encyclopaedia' at the beginning of Chapter 2.

Many of the themes of the novel tie in very closely to postcolonial theory, particularly the challenging of purity. Rushdie believes, like Bhabha, that purity gives too much power to the dominant form of culture:

Question: What is the opposite of faith?
Not disbelief. Too final, certain, closed. Itself a kind of belief.
Doubt.

Rushdie, 1988: 92

We can also see the celebration of hybridity in the work of various postcolonial musicians, usually labelled as 'world music'. It is perhaps unsurprising that music should be so quick to respond to cultural meetings. Unlike literature which is the preserve of the educated, music is something which is performed by all sorts of people across the world. The emergence of popular forms of music can be seen as an expression of the voice of the marginalised and so a very important source for postcolonial understandings of the world.

One of the best examples of a hybrid musical form emerging from the colonial experience is *raï*. The word *raï* is from the Arabic, meaning opinion. It emerged in Algeria as a way to express opinions and dissent. Since its beginnings in the early twentieth century it has been a political form of expression, so much so that shortly after independence in 1962, a number of prominent *raï* singers were arrested, and the ban on its radio broadcast was only lifted in 1985. *Raï* combines traditional North African and European musical traditions. It is sung in local dialect, thus separating it from both formal Arabic and the French spoken by the elites in the cities.

The original *raï* artists of the early twentieth century are now called '*shikh*' (for men) and '*shikha*' (for women, who were traditionally among the most important of the style's founders), and the new proponents of the form call themselves '*cheb*' or '*chebba*' to distinguish themselves. *Raï* rhythms have become increasingly eclectic, merging Arabic, African, flamenco, disco, hip-hop and reggae in recognition of the growing influence of global musical traditions. *Raï* is becoming an increasingly global form of musical expression but still holds onto its political roots. French-Algerians who are living in France will focus their lyrics on the problems and difficulties of being an immigrant.

Diasporaic (those people who have been displaced from their homes due to wars and colonialism) and migrant communities are also represented through music. We can see this in examples from rock (the mixing of the musical traditions of African slaves with European-American forms), to bangra (Punjabi traditional wedding musical traditions set to electronic techno beats), to contemporary global lounge (for example, the French musician Claude Challe, best known for his *Buddha Bar* collections).

Other postcolonial artists and novelists are more ambivalent about the nature of hybridity. In his book *Xala* (which he later directed as a film, the first feature film directed by an African), published in 1974, Senegalese author Sembène Ousmane examines the postcolonial condition of his country in the figure of one man, El Hadji Abdou Kader Beye. The book starts out with El Hadji as a successful member of the chamber of commerce and industry, led for the first time by a Senegalese man. On the same day, El Hadji is being married to his third wife. However, his happiness comes to an end on his wedding night when he is rendered impotent by a *xala*, a curse. The remainder of the story charts his attempts to reclaim his manhood with trips to local healers and *marabout*. His over-ambition in taking a third wife also over-stretches him in financial terms and, at the same time as he is having to deal with his inability to meet the responsibilities towards his three wives, his business dealings also fall apart. He is rendered bankrupt, two of his wives leave him, and he has to humiliate himself in front of a group of beggars in order to become a man again.

The story traces El Haji's attempts to be successful in both his 'westernised' life in which he is a businessman, wearing a European-style suit, speaking

Figure 7.1 Cover of *Xala*

French and drinking mineral water, and also as a polygamous Muslim whose status is measured in his ability to support multiple wives. These two facets of his personality – and perhaps by extension, the two aspects of contemporary Senegalese society – are shown to be intimately connected, as success in one sphere facilitates success in the other while the *xala* and the cost of his attempts to fix it soon start to undermine his standing in the business sphere. Consider the following extracts (Ousmane, 1974: 2–3, 16, 17–18, 47), and think about the ways in which the different facets of El Hadji's character are interdependent. Also think about why this particular form of curse was placed on him. Might there be connections with Fanon's 'cure' for colonialism?

> The [Business] Group's President paused in his speech. His eyes shone with sat-isfaction as they came to rest on each member of his audience in turn: ten or so expensively dressed men. The cut of the made-to-measure suits and their immaculate shirts were ample evidence of their success. [...]

'We are the leading businessmen in the country, so we have a great responsibility. … But it is time now to bring this memorable day to a close by reminding you that we are invited to the wedding of our colleague El Hadji Abdou Kader Beye. Although we are anxious to belong to the modern world we haven't abandoned our African customs. I call upon El Hadji to speak.'

El Hadji Abdou Kader Beye, who was seated on the President's right, rose to his feet. His close-cropped hair was streaked with white but he carried his fifty-odd years well.

'Friends, at this precise moment (looking at his gold wrist-watch) the marriage has been sealed at the mosque. I am therefore married. … I have now married my third wife, so I'm a 'captain' as we Africans say. Mr. President, will you all do me the honour of being my guests?'

'A fitting way to end the day. Gentlemen, the women are waiting for us. Shall we go?'

The meeting was over.

Outside a line of expensive cars was waiting for them. El Hadji Abdou Kader Beye drew the President to one side: 'Take the head of the convoy. I must go and collect my other two wives.' […]

As they came out of the front door Oumi N'Doye [El Hadji's second wife] whispered to El Hadji:

'Which of us is to sit in the back with you?'

Before El Hadji could reply she continued: 'All three of us then. After all it isn't her *moomé* [the period a polygamist spends with each of his wives].' […]

[The bridal chamber] was decorated completely in white. A mattress laid on the floor in a corner, an upside-down mortar and a woodcutter's axe-handle were for the moment the only furnishings.

'It is time for you to change, El Hadji,' the Bayden told the man.

'Change? What for?'

'You must put on a caftan without trousers and sit on the mortar, with the axe-handle between your feet, until your wife's arrival is announced.'

'Yay, Bineta, you don't really believe in all that! I have two wives already and I did not make a fool of myself with this hocus-pocus on their account. And I am not going to start today!'

'You're not a European, although I can't help wondering.'

[… once the *xala* has taken effect, El Hadji is more willing to seek help in the African traditions he ridiculed on his wedding night…]

El Hadji Abdou Kader Beye followed the Marabouts' instructions: he drank their concoctions, rubbed himself with their ointments and wore their *xatim* round his waist. … Day after day, night after night, his torment ate into his professional life. Like a waterlogged silk-cotton tree on the river bank he sank deeper into the mud. Because of his condition he avoided the company of his fellow business-men, among whom transactions were made and sealed. He was weighed down by worry and lost his skill and his ability to do business. Imperceptibly his affairs began to go to pieces.

He had to maintain his high standard of living: three villas, several cars, his wives, children, servants and employees. Accustomed to settling everything by cheque, he continued to pay his accounts and his household expenses in this way. He went on spending. Soon his liabilities outstripped his credit.

FUNDAMENTALISM

As we saw in the previous chapter, fundamentalism offers resistance to the perceived homogenisation of culture which is sometimes interpreted as a new colonisation of western goals, values and beliefs through the media. Thus, at the same time as the emergence of hybrid cultures, global identities and cultural productions, we are also seeing new forms of reactionary iden-tification emerge.

It is important always to remember that novels are not just collections of words, they are also real, physical objects which have a life in the real world and, as subsequent events proved for Salman Rushdie, ideas contained within the pages of books do not necessarily stay there. Cultural productions 'escape' the place they were produced in by becoming part of the global media, the global market, and the global culture, published in many languages simulta-neously. Authors cannot control the effects and interpretations of their work, and in the case of the *The Satanic Verses* there are limits to this fluidity demonstrating Massey's power-geometries. Despite the fluidity of globalisa-tion and hybridity, Rushdie discovered after the publication of *The Satanic Verses* that there were still real boundaries and real identities in play.

We can see this in reactions to this book and works by others who chal-lenged religious/traditional authority. The *Satanic Verses* had a number of offending passages which challenged the purity of the Qu'ran as a religious text. The problem was not just that someone questioned the purity of the Qu'ran but that it was done by Rushdie, a man heralded by the western press as having written the ultimate South Asian novel *Midnight's Children* (about the formation of post-colonial India), a man who had been educated as a Muslim but had turned against the faith in this most secular of productions. Ayatollah Khomeini, the Iranian cleric, put a *fatwa* (a religious death penalty) on Rushdie for his blasphemy. There were marches, protests and book-burnings in the Middle East and among Muslim communities in the west. The author in turn had to go into hiding.

Rushdie thereby lost control of the interpretation of *The Satanic Verses*. This is testament to the importance of literature in the construction of 'reality', and the inseparability of 'fictional' texts and those which purport, more directly, to represent reality. For Rushdie, the division between fact and fiction became highly permeable. His death under the Islamic *fatwa* was fictionalised by Brian Clark in a play entitled, *Who killed Salman Rushdie?* Meanwhile, other elements of the media opened up his life to widespread debate:

> On TV shows, studio audiences were asked for a show of hands on the question of whether I should die. A man's murder (mine) became a legitimate subject for a national opinion poll. (Rushdie, 1991: 407)

The various reactions to *The Satanic Verses* (book-burning in Bradford, the violent demonstrations in Pakistan and India, and the *fatwa* itself) could have come straight from a section of Rushdie's own writing about the hybrid and media-influenced world in his novel.

It is therefore an irony that Rushdie's writing of *The Satanic Verses* has erased his authorship of it. Of the reams of print devoted to *The Satanic Verses*, only a tiny proportion has discussed the novel itself. Spivak argued that 'it is the late Ayatollah who can be seen as filling the Author-function, and Salman Rushdie, himself, caught in a different cultural logic, is no more than the writer-as-performer'(Spivak, 1990: 42).

The literary mode that Rushdie employs is also revealing of the nature of the extreme reactions to it. He has written the majority of his work as novels. Thus, Rushdie has chosen a very specific literary convention, associated with a particular period of western history. Whilst adopting this form of writing, however, Rushdie has not been faithful to its conventions. Instead, he has subverted the novelistic style, forcing it to represent the hybridity that his protagonists display. As a post-colonial subject, Rushdie recolonises this western mode of representation to tell of the ambiguity, rather than the singularity, of subjectivity. The structure of the text therefore mirrors the structure of the world that Rushdie has created.

> The use of fiction was a way of creating the sort of distance from actuality that I felt would prevent offense. I was wrong. (Rushdie, 1991: 408–9)

According to Said, for fundamentalist interpretations of Islam novels representing alternative representations of the world are deeply problematic. The novel, 'an institutionalization of the intention to begin' (Said, 1985: 100), is especially challenging to this worldview:

> … it is significant that the desire to create an alternative world, to modify or augment the real world through the act of writing (which is one motive underlying the novelistic tradition in the West) is inimical to the Islamic world-view. The Prophet is he who has *completed* a world-view; thus the word *heresy* in Arabic is synonymous with the verb 'to innovate' or 'to begin.' Islam views the world as a plenum, capable of neither diminishment nor amplification. (Said, 1985: 81)

The Satanic Verses presents an alternative account of Mohammed's writing of the Qu'ran. Although this passage occurs in the dream sequence of a character whose religious doubt ultimately dooms him, it is also written within a novel which questions the distinction between good and bad, sacred and profane. The religious purity of the Qu'ranic verses cannot be questioned, especially by an author heralded by the literary institutions of the west as the saviour of Southern Asian writing. The subsequent award to Rushdie in 2007 of a knighthood by the British state again saw the emergence of renewed threats in the context of post-2001 western-Islamic tensions.

In the west, the 'Rushdie affair' was often interpreted as arising from the parochial, even arcane, beliefs of certain Islamic leaders, which ran against the modern, universal value of freedom of speech. However, it would be disingenuous to suggest that it was only those Islamic voices who represented a particular belief that was set in this debate:

> ... there is another absurdity, equally extreme, equally common. And that is to pretend that the scriptor/author is ignorant of the social context in which the work is produced, the potential for a novel that draws, *however sympathetically*, on the life of the prophet to become an orientalist icon in a racist world ... *The Satanic Verses*, regardless of the author's inclinations, is marked by the stigmata of the literary world that defines it as a classic. (Keith and Pile, 1993: 33)

While there was a great deal of global media coverage of Muslims burning copies of Salman Rushdie's book for being heretical, for instance, and thereby reinforcing a popular image that linked fundamentalism with Islam, there was very little coverage of US Christian fundamentalists burning J.K. Rowling's *Harry Potter* books for promoting witchcraft.

This again highlights the fact that globalisation is clearly not negating geography but making it more complex and ambivalent. This fluidity is in tension with the material geographies of borders, nationalisms and religions. It is also important to consider the language and examples used in these kinds of debates. Some have pointed to the actuality of borders much more clearly and directly. Resonating with Massey's ideas about power-geometries, Chicana writer Gloria Anzaldúa writes about the nature of the border between the USA and Mexico. Like Rushdie, she combines languages and a mix of fact and fiction in her writing to conjure up an image of the lived realities that the border cuts through:

> To live in the Borderlands means you
>
> > are neither *hispana india negra española*
> > *ni gabacha, eres mestiza, mulata,* half-breed
> > caught in the crossfire between camps
> > while carrying all five races on your back
> > not knowing which side to turn to, run from;
>
> To live in the Borderlands means knowing

that the india in you, betrayed for 500 years,
is no longer speaking to you,
that *mexicanas* call you *rajetas*,
that denying the Anglo inside you
is as bad as having denied the Indian or Black;

Cuando vives en la frontera

people walk through you, the wind steals your voice,
you're a *burra, buey*, scapegoat,
forerunner of a new race,
half and half – both woman and man, neither –
a new gender;

To live in the Borderlands means to

put *chile* in the borscht,
eat whole wheat *tortillas*,
speak Tex-Mex with a Brooklyn accent;
be stopped by *la migra* at the border checkpoints;

Living in the Borderlands means you fight hard to

resist the gold elixir beckoning from the bottle,
the putt of the gun barrel,
the rope crushing the hollow of your throat;

In the Borderlands

you are the battleground
where enemies are kin to each other;
you are at home, a stranger,
the border disputes have been settled
the volley of shots have shattered the truce
you are wounded, lost in action
dead, fighting back;

To live in the Borderlands means

the mill with the razor white teeth wants to shred off
your olive-red skin, crush out the kernel, your heart
pound you pinch you roll you out
smelling like white bread but dead;

To survive the Borderlands

you must live *sin fronteras*
be a crossroads.

. . .

1,950 mile-long open wound

> dividing a *pueblo*, a culture,
> running down the length of my body,
> staking fence rods in my flesh,
> splits me splits me
> *me raja me raja*

> This is my home
> this thin edge of
> barbwire.

>   …

The U.S.-Mexican border *es una herida abierta* where the Third World grates against the first and bleeds.

Gloria Anzaldúa, 1987, *Borderlands/La Frontera: The New Mestiza*, pp.194–5, 2–3

Anzaldúa is quite specific about why she is writing in this way, and it is similar to Rushdie's use of language while also clearly echoing Spivaks politics of voice we came across in the previous chapter:

> The switching of 'codes' in this book from English to Castillian Spanish to the North Mexican dialect to Tex-Mex to a sprinking of Nahuatl to a mixture of all of these, reflects my language, a new language – the language of the Borderlands. There, at the juncture of cultures, languages cross-pollinate and are revitalized; they die and are born. Presently this infant language, this bastard language, Chicano Spanish, is not approved by any society. But we Chicanos no longer feel that we need to beg entrance, that we need always to make the first overture – to translate to Anglos, Mexicans and Latinos, apology blurting out of our mouths with every step. Today we ask to be met halfway. (Anzaldúa, 1987: viii)

Further Reading

The best reading for this section are the novels already mentioned. Others you could read include Hanif Kureishi's *The Buddha of Suburbia* (1991) and Monica Ali's *Brick Lane* (2003).

On discussions of Third World Literature

Ahmad, A. (1987) 'Jameson's rhetoric of Otherness and the "national allegory"', *Social Text*, 17: 3–25.
Jameson, F. (1986) 'Third World literature in an age of multinational capitalism' *Social Text*, 15: 65–88.
Prasad, M. (1992) 'On the question of a theory of (third world) literature', *Social Text*, 31/32: 57–83.

(Cont'd)

On The Satanic Verses

Sharp, J. (1996) 'Locating imaginary homelands: literature, geography and Salman Rushdie', *GeoJournal*, 38 (1): 119–27.
Spivak, G. (1990). 'Reading *The Satanic Verses*. *Third Text*, 11: 41–60.

On cinema

Downing, J. (ed.) (1987) *Film and Politics in the Third World*. New York: Autonmedia.
Nwachukwu, F. (1994) *Black African Cinema*. London: University of California Press.
Shohat, E. (1997) 'Post-Third-Worldist culture: gender, nation and the cinema', in J. Alexander and C. Mohanty (eds), *Feminist Genealogies, Colonial Legacies, Democratic Futures*. London: Routledge.

8 *LEAVING THE ARMCHAIR?*

Across the various examples in this book, we have looked at the arguments put forward by postcolonial theory and considered what geographies the colonial and post-colonial processes have generated. Postcolonial approaches have been incredibly influential in recent years, not only in the literary and cultural fields within which they first emerged, but also in disciplines across the social sciences and arts. As is the case for any approach that is influential, postcolonialism has come in for much critique. Although we have implicitly dealt with many of these critiques as we have worked through these chapters, in this final one we will consider a single critique directly to allow for an evaluation of postcolonialism and postcolonial geography as approaches to understanding relations between 'them' and 'us' today.

POSTCOLONIALISM AND DEVELOPMENT

The main critique that has been levelled at postcolonialism as an intellectual project has been its concentration on texts and on representation. We have come across these ideas already, and it has been the aim of this book to demonstrate the advantages of a specifically *geographical* approach to postcolonialism. For instance, we have seen how postcolonial landscape geographies have necessitated going beyond the ideal representations of space in maps and plans once drawn up for particular places, to understand the meanings that arise through day-to-day use of landscapes.

Nevertheless, in a related and powerful critique, Gross (1996: 248) has argued that postcolonial critics, 'have guaranteed themselves the position of armchair decolonisers', by which he means that the ideas of postcolonialism have been generated from inside the west, inside academia, and from an analysis of texts rather than fieldwork. The clever ideas of postcolonial theory are sometimes challenged for being too caught up with producing critiques of the texts central to western thought (whether western philosophy, literature or art) and for not spending enough time considering the real issues being faced by people in the global south today. For instance, some would wonder as to what the cultural and theoretical bases of postcolonialism can tell us about poverty, inequality, racism, subjugation … Indeed, a number of commentators have noted that the impact of postcolonialism on the study of geography has been a move away from research outside of the west. Postcolonial theories about representation have meant that some have felt too anxious about the possibilities of their own misrepresentation of other places and peoples and therefore have sought to examine other people's accounts instead. It seems that, as geographers, we are presented with a dilemma which Robert Young puts like this: 'If you participate [in research] you are, as it were, an Orientalist, but of course if you don't then you're a eurocentrist ignoring the problem' (quoted in Sparke, 1994: 119).

These critiques of postcolonialism are particularly associated with those for whom colonialism and its aftereffects were (and are) primarily – and most fundamentally – about economic exploitation and dominance. We can see this played out by radical development theorists particularly, who see the condition of countries in the global south today as resulting from economic exploitation. Taking this view, the cultural focus of postcolonialism is regarded as trivial compared to what is regarded as the real issues of poverty, hunger and disease.

While the value of development theorists' focus on the material conditions of people's survival is, of course, impossible to deny, I would want to suggest here that this does not negate the power of postcolonial arguments. As we have seen throughout this book, the cultural is never simply superficial: fictional accounts do not only reside within the covers of books but spill out into the real world; into people's expectations of new places and their reactions to these places; into academic accounts of foreign societies informing political practices, whether of colonial rule in the nineteenth century or the 'war on terror' in the twenty-first; into diasporic music as not only a collection of tones and words but as real connections between people stretched across the globalised geographies of home that characterise the contemporary world.

Moreover, postcolonial theorists have highlighted the continued use of colonial conceptions of the world by development theory and practice. For instance, the notion of 'less developed' or 'developing' countries is clearly still very much imbued with the notion of an evolutionary progression of societies characteristic of nineteenth century colonial mindsets, albeit expressed in a different language. And of course, what we cannot escape is that, despite the initial optimism, and despite our expectations of the 'development decade' and the passage of several subsequent decades, poverty has not been eradicated – rather, the poor are becoming poorer for development has not worked.

Relations between development theory and postcolonialism have been nothing short of frosty. As a result of this mutual distrust, there has been little in the way of intellectual exchange between development theorists and practitioners and postcolonialism. As Christine Sylvester (1999: 703) has put it, 'development studies does not tend to listen to subalterns and postcolonial studies does not tend to concern itself with whether the subaltern is eating.'

And yet, as Sylvester herself has argued, there is much to be gained from closer communication between development theory and postcolonialism: in terms of grounding postcolonialism in the real life struggles for survival of poor people around the world, and of making development more conscious of the concepts it relies upon. One example of how this has been taking place is in the area of indigenous knowledges.

Some theorists have suggested that the failure of development is in large part because of the failure of development practitioners to listen to the knowledge of those peoples to be developed. Development is done to people in the poor countries of the world by people from the wealthier areas. Development practices are drawn from theories which privilege western knowledge – and particularly scientific knowledge – as the form of knowledge that will facilitate development. Some have turned to 'indigenous knowledges', the ways of knowing that are held by poor people, as a way of making development more relevant and appropriate, and therefore, hopefully, more successful. After all, although poor these people have developed cultural values that cannot simply be ignored and ways of working with often harsh environments from which western 'experts' can undoubtedly learn. It is therefore clear how this approach relates to the postcolonial concern with listening to others.

As we have seen throughout this book, postcolonial approaches show the complexity of the world and, even more importantly, emphasise the importance of respecting other voices and their explanations, goals and aspirations. However, if indigenous knowledges are to be genuinely brought into conversation with western notions of development, this does have to be a true exchange and cannot be a simple case of incorporation. Western development as a knowledge must be open to change, however difficult this might be. In discussing scientists' fears that fully embracing the significance of indigenous knowledge might lead to the validation of approaches such as creationism and astrology, Nakashima and de Guchteneire (1999: 40) suggest that:

we might consider that the discomfort of these scientists gives expression to a more fundamental concern … about the relationship between science and these other systems of knowledge, other understandings of the world. Of course, if indigenous knowledge is conceived as just another information set from which data can be extracted to plug into scientific frameworks of understanding, then we do not trouble the scientific worldview. However, this practical approach – that of the pharmaceutical industry or of conservation ecologists who validate traditional information and use it to attain pre-defined ends – may threaten the integrity of traditional knowledge systems. On the other hand, if science is seen as one knowledge system among many, then scientists must reflect on the relativity of their knowledge and their interpretations of 'reality'. For the survival of traditional knowledge as a dynamic, living and culturally meaningful system, this debate cannot be avoided.

To follow through the implications of postcolonialism for development, then, would mean a decentring of western science and other western theories used for development. Postcolonialism shows development studies the importance of representations on the way people act. But putting this into practice is very difficult: what happens if the indigenous knowledges are contradictory to the development approach? Whose knowledge should be listened to (the elders, the 'head of household', women)? How should conflicting views be negotiated? These are very challenging issues, but they must be addressed to ensure that any development programmes are appropriate to the peoples they are aimed at.

BRINGING DIFFERENT VOICES INTO DEVELOPMENT: A CASE STUDY FROM EGYPT

In 1989 Wadi Allaqi, a valley running into the high dam lake in the south of Egypt, was declared a Protected Area under Egyptian law, and subsequently as a UNESCO Biosphere Reserve in 1994. As part of the process, those Bedouin communities resident within the area were consulted about the proposals. However, the key decisions were taken within the context of a western environmental discourse which was very different from the understanding of the environment held by the Wadi Allaqi Bedouin. In line with western environmental practice, boundaries were drawn around different tracts of land, to produce core and buffer zones, which were to have different degrees of environmental protection. Within these boundaries there were particular conservation practices that were required to be legally observed. For Bedouin however, the idea of drawing boundaries around areas of land was alien. Resources are defined by Bedouin in a more fluid manner. Conservation reflects both community needs and differing drought pressures on different vegetation resources at particular times, both on annual and significantly longer timescales. Conservation is a temporal practice for them, necessary at certain times of the year or in particular seasons. This cyclical, temporal knowledge of conservation and resources is different to a western spatial definition which constrains or excludes certain practices in defined geographic locations.

The difference in knowledges of environmental management is clear, for instance in the example of conflicting understandings about the conservation of acacia trees by Bedouin and Western conservationists. Acacia trees constitute a centrally important economic resource for the Bedouin and are exploited in a sustainable way. They provide a source of feed for livestock from naturally fallen leaves, shaken leaves and fruit. They also provide an important source of wood for charcoal making; acacia is particularly valued for the quality of charcoal that can be made from it. Access to the various economic elements of acacia tress and bushes can be complex. From the same tree, one family may have claims to only naturally fallen leaves, while another family may have access to those leaves which are dislodged when the plant is shaken, and a third family to only the dead wood for charcoaling. For another tree, one family may have rights to all of its production. The situation can be further complicated by the existence of some prohibitions against taking resources during some times in the year, whereas at other times resources can be removed without any such difficulty. This method, therefore, provides for a system of conservation of scarce resources, even though this may not necessarily meet the requirements of formal, western-based, conservation practice. Bedouin conceptualisations of conservation are culturally and economically embedded, and managed within in the interests of their wider community interests.

Figure 8.1 Bedouin women shaking the branches of a tree to dislodge leaves for their goats

Prohibition on removing acacia in the new environmental area led to further problems. In 1998, water in the High Dam lake rose to unprecedented levels. As a result, about 12 mature acacias were inundated and died. In these circumstances, Bedouin would traditionally use the trees to make charcoal due to the fact that they would never again produce new wood. However, as the trees had grown within the conservation area, there was a prohibition against their use by humans. And so, in order to comply with the regulations imposed by the conservation area label, the Bedouin were expected to ignore the dead trees. Unsurprisingly, the Bedouin saw little logic in the formal, western

(Cont'd)

position of such conservation. There was a clear cultural divide between the two rather different views of conservation.

To ensure a programme of development – in this case, articulated through a conservation scheme – that will work in the long term, the views and understandings of all those who live in the area need to be considered. Indeed, the Bedouins' flexible system of conservation might prove a very important alternative strategy to more conventional approaches.

Taken from Briggs and Sharp (2004)

CONCLUSION

What *geographies* of postcolonialism have shown us is that trying to see post-colonialism, and material or economic accounts of the colonial and post-colonial periods, as competing explanations, is untenable. The ways that material issues are conceived and acted upon cannot be disentangled from inherited cultural values and knowledges, while postcolonial cultural representations and knowledges have profoundly material consequences. Culture is neither a luxury nor a trivial diversion, but is central to the imagining and (re)making of the world around us. The power of postcolonialism allows us to look beyond our inherited geographical imaginations and makes it possible for us to start imagining new ones.

Further reading

On postcolonialism, development and indigenous knowledges

Gross, J. (1996) 'Postcolonialism: subverting whose empire?', *Third World Quarterly*, 17 (2): 239–50.

Nakashima, D. and de Guchteneire, P. (1999) 'A new impetus for indigenous knowledge from the World Conference on Science', *Indigenous Knowledge and Development Monitor*, 7 (3): 40.

Sylvester, C. (1999) 'Development studies and postcolonial studies: disparate tales of the "Third World"', *Third World Quarterly*, 29 (4): 703–21.

On the conflict between indigenous knowledges and ideas of development

Briggs, J. and Sharp, J. (2004) 'Indigenous knowledges and development: a postcolonial caution', *Third World Quarterly*, 25 (4): 661–76.

Silvern, S. (1995) 'Nature, territory and identity in the Wisconsin treaty rights controversy', *Ecumene*, 2 (3): 267–92.

BIBLIOGRAPHY

Abu-Lughod, L. (1995) 'The objects of soap opera: Egyptian television and the cultural politics of modernity', in D. Miller (ed.), *Worlds Apart: Modernity Through the Prism of the Local*. London: Routledge. pp. 190–210.

Adas, M. (1989) *Machines as the Measures of Man*. Ithaca, NY: Cornell University Press.

Ahmad, A. (1987) 'Jameson's rhetoric of Otherness and the "national allegory"', *Social Text*, 17: 3–25.

Ahmad, A. (1992) 'Orientalism and after', in A. Ahmad, *In Theory*. London: Verso. (Reprinted in P. Williams and L. Chrisman (eds) (1994), *Colonial Discourse and Post-colonial Theory*. New York: Columbia. pp. 162–71.)

Ali, M. (2003) *Brick Lane*. London: Doubleday.

Allen, R.C. (1995) *To Be Continued: Soap Operas Around the World*. London: Routledge.

Anzaldúa, G. (1987) *Borderlands/La Frontera: The New Mestiza*. San Francisco, CA: Aunt Lute.

Appadurai, A. (1994) 'Disjuncture and difference in the global cultural economy', in P. Williams, and L. Chrisman (eds), *Colonial Discourse and Post-colonial Theory*. New York: Columbia. pp. 324–39.

Arnold, D. (1993) *Colonizing the Body*. Berkeley, CA: University of California Press.

Aziz (2004) Viewpoint: why I decided to wear the veil. *BBCi news*. http//:www/bbc.co.uk/go/ pr/fr/-/1/hi/talking_point/3110368.stm. viewed 3/2/04.

Barber, B. (1992) 'Jihad vs McWorld', *The Atlantic Magazine,* March.

Barthes, R. (1956) *Mythologies*. London: Grant and Culter.

Baudet, H. (1988) *Paradise on Earth: Some Thoughts on European Images of Non-European Man* (trans. E. Wentholt). Middletown, CT: Wesleyan Press.

Bhabha, H. (1990) 'Novel metropolis', *New Statesman and Society*, 16 Feb: 16–18.

Bhabha, H. (1994) *The Location of Culture*. London: Routledge.

Blunt, A. (1994) *Travel, Gender and Imperialism: Mary Kingsley and West Africa*. New York: Guilford Press.

Blunt, A. (2005) *Domicile and Diaspora: Anglo-Indian Women and the Spatial Politics of Home*. Oxford: Blackwell.

Blunt, A. and McEwan, C. (2002) *Postcolonial Geographies*. London: Continuum.

Briggs, J. and Sharp, J. (2004) 'Indigenous knowledges and development: a postcolonial caution', *Third World Quarterly*, 25 (4): 661–76.

Bruner, E. (1991) 'Transformation of self in tourism', *Annals of Tourism Research*, 18: 238–50.

Carter, P. (1987) *The Road to Botany Bay: An Exploration of Landscape and History*. New York: Knopf.

Chatterjee, P. (1986) *Nationalist Thought and the Colonial World: A Derivative Discourse*. Minneapolis: University of Minnesota.

Chatterjee, P. (1993) *The Nation and Its Fragments*. Princeton, NJ: Princeton University Press. (See especially the chapter 'Whose imagined community?'.)

Childs, P. and Williams, P. (1997) *An Introduction to Post-colonial Theory*. London: Prentice Hall.

Conrad, J. (1988 [1926]) *Heart of Darkness* (ed. R. Kimbrough) 3rd edn. London and New York. W.W. Norton and Co.

Coombes, A. (1994) *Reinventing Africa: Museums, Material Culture and Popular Imagination*. New Haven: Yale University Press.

Currier, G. (1832) *The Animal Kingdom: Arranged in Conformity With its Organization.* Abridged version. New York: G. & C. & H. Carrill.

Darwin, C. (1859) *On the Origin of Species by Natural Selection, or the Preservation of Favoured Races in the Struggle for life.* London: Murray.

Desai, K. (2006) *The Inheritance of Loss.* London: Hamish Hamilton Ltd.

Domosh, M. (1991) 'Towards a feminist historiography of geography', *Transactions of the Institute of British Geographers,* 16: 95–104.

Downing, J. (ed.) (1987) *Film and Politics in the Third World.* New York: Autonmedia.

Driver, F. (1991) 'Henry Morton Stanley and his critics: geography, exploration and empire', *Past and Present,* 133: 134–66.

Driver, F. (2001) *Geography Militant: Cultures of Exploration and Empire.* Oxford: Blackwell.

Duncan, J. (1990) *The City as Text.* Cambridge: Cambridge University Press.

Duncan, J. (1992) 'Re-presenting the landscape: problems of reading the intertextual,' in L. Mondada, F. Panese, and O. Söderström, (eds), *Paysage e crise de la lisibilité.* Lausanne: Université de Lausanne, Institut de Géographie, pp. 81–93.

Duncan, J. (2002) 'Embodying colonialism?: Domination and resistance in 19th century Ceylonese coffee plantations', *Journal of Historical Geography,* 28, (3): 317–38.

Duncan, J. (2007) *In the Shadows of the Tropics: Climate, Race and Biopower in Ninteenth Century Ceylon.* Aldershot: Ashgate.

Fanon, F. (1963) 'National Culture', in *The Wretched of the Earth.* New York: Grove Weidenfeld, pp. 206–48.

Fanon, F. (1963) *The Wretched of the Earth.* New York: Grove Weidenfeld. Especially 'Concerning violence'

Fanon, F. (1967) *Black Skin, White Masks.* New York: Grove Press.

Foucault, M. (1970) *The Order of Things: An Archaeology of the Human Sciences.* London: Tavistock.

Frankenberg, R. (1993) *White Women, Race Matters: The Social Construction of Whiteness.* Minneapolis: University of Minnesota Press.

Friedman, J. (1981) *The Monstrous Races in Medieval Art and Thought.* Harvard: Harvard University Press.

Godlewska, A. (1994) 'Napoleon's geographers (1797–1815): imperialists and soldiers of modernity', in A. Godlewska and N. Smith (eds), *Geography and Empire.* Oxford: Blackwell.

Greenhalgh, P. (1988) *Ephemeral Vistas: The Expositions Universelles, Great Exhibitions and World's Fairs, 1851–1939.* Manchester: Manchester University Press.

Gregory, D. (2004) *The Colonial Present.* Oxford: Blackwell.

Gross, J. (1996) 'Postcolonialism: subverting whose empire?', *Third World Quarterly,* 17 (2): 239–50.

hooks, b. (1990) 'Marginality as a site of resistence', in R. Ferguson et al. (eds), *Out There: Marginalization and Contemporary Cultures.* Cambridge, MA: MIT. pp. 341–43.

Huggins, J., Huggins, R. and Jacobs, J. (1995) 'Kooramindanjie: place and the postcolonial', *History Workshop Journal,* 39: 165–81.

Huntington, E. (1907) *The Pulse of Asia: A Journey in Central Asia Illustrating the Geographic Basics of History.* London: Constable.

Huntington, E. (1915) *Civilization and Climate.* New Haven, CT: Yale University Press.

Huntington, S. (1993) 'The clash of civilizations' *Foreign Affairs,* 72 (3): 22–49.

Irvine, R. (1981) *Indian Summer: Lutyens, Baker and Imperial Delhi.* London: Yale University Press.

Iyer, P. (1988) *Video Night in Katmandu: And Other Reports from the Not-So-Far East.* New York: Vintage.

Jacobs, J. (1996) *Edge of Empire.* London: Routledge.

Jameson, F. (1986) 'Third World literature in an age of multinational capitalism', *Social Text,* 15: 65–88.

Keith, W. and Pile, S. (1993) 'Introduction part 2: The place of politics', in W. Keith and S. Pile (eds), *Place and the Politics of Identity*. New York: Routledge, 22–40.

Kenny, J. (1995) 'Climate, race, and imperial authority: the symbolic landscape of the British hill station in India', *Annals, Association of American Geographers*, 85: 694–714.

King, A.D. (1976) *Colonial Urban Development*. London: Routledge.

Kingsley, M. (1897) *Travels in West Africa*. London: MacMillan.

Klein, N. (2007) 'Using Crisis to Take on the Fakes', *The Herald*, Oct 4, p. 7.

Kumar, S. (2002) 'The evolution of spatial ordering in colonial Madras', in A. Blunt and C. McEwan (eds), *Postcolonial Geographies*. London: Continuum.

Kureshi, H. *The Buddha of Surburbia*. London: Continuum International Publishing Group.

Lefebvre, H. (1991) *The Production of Space*. Oxford: Blackwell.

Lévy, B-H. (2004), 'Off with their headscarves', *The Sunday Times*, 1/2/04, p. 5.2.

Loomba, A. (1998) *Colonialism/Postcolonialism*. London: Routledge.

Lutz, C. and Collins, J. (1993) *Reading National Geographic*. Chicago: University of Chicago Press.

Massey, D. (1993) 'Power-geometry and a progressive sense of place', in J. Bird, B. Curtis, T. Putnam and G. Robertson (eds), *Mapping the Futures*. London: Routledge.

MacCannell, D. (1976) *The Tourist: A New Theory of the Leisure Class*. New York: Schocken.

MacCannell, D. (1992) *Empty Meeting Grounds*. London: Routledge.

McClintock, A. (1995) *Imperial Leather*. London: Routledge.

MacKenzie, J. (ed.) (1986) *Imperialism and Popular Culture*. Manchester: Manchester University Press.

Mitchell, K. (1997) 'Different diasporas and the hype of hybridity', *Environment and Planning D: Society and Space*, 15: 533–53.

Mitchell, T. (1988) *Colonising Egypt*. Berkeley, CA: University of California Press.

Moaveni, A. (2005) *Lipstick Jihad: A Memoir of Growing Up Iranian in America and American in Iran*. Cambridge, MA: Perseus Books Group.

Mohanty, C., Russo, A. and Torres L. (eds) (1991) *Third World Women and the Politics of Feminism*. Bloomington, IN: University of Indiana Press.

Myers, G. (2003) *Verandas of Power: Colonialism and Space in Urban Africa*. Syracuse, NY: Syracuse University Press.

Nakashima, D. and de Guchteneire P. (1999) 'A new impetus for indigenous knowledge from the World Conference on Science', *Indigenous Knowledge and Development Monitor*, 7 (3): 40.

New Internationalist (2005) 'The unreported year 2004', *New Internationalist*, p. 8–9.

Nwachukwu, F. (1994) *Black African Cinema*. London: University of California Press.

Ouseman, S. (1974) *Xalae* Chicago: Chicago Review Press.

Peet, R. (1985) 'The social origins of environmental determinism', *Annals of the Association of American Geographers*, 75: 309–33.

Philips, R. (1997) *Mapping Men and Empire*. London: Routledge.

Pletsch, C. (1981) 'The three worlds, or the division of social scientific labor, circa 1950–1975', *Comparative Studies in Society and History*, 565–590.

Porter, D. (1994) 'Orientalism and its problems', in P. Williams and L. Chrisman (eds), *Colonial Discourse and Post-colonial Theory*. New York: Columbia. pp. 150–61.

Prasad, M. (1992) 'On the question of a theory of (third world) literature', *Social Text*, 31/32: 57–83.

Pratt, M.L. (1992) *Imperial Eyes: Travel Writing and Transculturation*. London: Routledge.

Radcliffe, S. (1994) '(Representing) post-colonial women: authority, difference and feminisms', *Area*, 26: 25–32.

Riffenburgh, B. (1993) *The Myth of the Explorer*. London: Wiley.

Rostow, W. (1960) *Stages of Economic Growth: A Non-communist Manifesto*. Cambridge: Cambridge University Press.

Rothenberg, T. (1994) 'Voyeurs of imperialism: the *National Geographic Magazine* before World War II', in A. Godlewska and N. Smith (eds), *Geography and Empire*. Oxford: Blackwell.

Roy, A. (2002) 'Come September', *Zmag.* http://www.zmag.org/content/print_article.cfm? itemID2404&sectionID=15 (accessed 1st March 2004).

Rushdie, S. (1981) *Midnight's Children.* London: Picador.

Rushdie, S. (1982) 'The empire writes back with a vengeance', *The Times,* July 3, p. 10.

Rushdie, S. (1988) *The Satanic Verses.* New York: Viking.

Rushdie, S. (1991) *Imaginary Homelands: Essays and Criticism, 1981–1991.* New York: Granta.

Ryan, S. (1996) *The Cartographic Eye: How Explorers Saw Australia.* Cambridge: Cambridge University Press.

Said, E. (1978) *Orientalism.* New York: Vintage.

Said, E. (1985) 'Orientalism reconsidered', *Cultural Critique,* 1: 89–107.

Said, E. (1993) *Culture and Imperialism.* New York: Knopf.

Said, E. (1999) *Out of place: a memoir.* New York: Vintage.

Sartre, J-P. (1963) 'Preface', in F. Fanon (ed.), *The Wretched of the Earth.* New York: Grove Weidenfeld.

Shohat, E. (1997) 'Post-Third-Worldist culture: gender, nation and the cinema', in Alexander et al. (eds), *Feminist Genealogies, Colonial Legacies, Democratic Futures.* London: Routledge.

Shohat, E. and Stam, R. (1994) *Unthinking Eurocentrism.* London: Routledge.

Sidaway, J. (2000) 'Postcolonial geographies: an exploratory essay', *Progress in Human Geography,* 24 (4): 591–612.

Silvern, S. (1995) 'Nature, territory and identity in the Wisconsin treaty rights controversy', *Ecumene,* 2 (3): 267–92.

Smith, P. (1988) 'Visiting the Banana Republic', in A. Ross (ed.), *Universal Abandon?* Minneapolis: Minnesota University Press.

Sparke, M. (1994) 'White mythologies and anemic geographies: a review', *Environment and Planning D: Society and Space,* 12: 105–23.

Sparke, M. (2005) *In the Space of Theory: Postfoundational Geographies of the Nation-State.* Minneapolis: University of Minnesota Press.

Spivak, G. (1990) Reading *The Santanic Verses. Third Text,* 11: 41–60.

Spivak, G.C. (1994) 'Can the Subaltern Speak?', in P. Williams and L. Chrisman (eds), *Colonial Discourse and Post-colonial Theory.* New York: Columbia. pp. 66–111.

Stanley, H.M. (1878) Through the dark continent: or, the sources of the Nile around the great lakes of equatoral Africa and down the Livingstone River to the Atlantic Ocean. London: Sampson, Low, Marston, Searle and Rivington.

Stoddart, D. (1986) *On Geography.* Oxford: Blackwell.

Stoddart, D. (1991) 'Do we need a feminist historiography of geography – and if we do, what should it be like?', *Transactions of the Institute of British Geographers,* 16: 484–87.

Sylvester, C. (1999) 'Development studies and postcolonial studies: disparate tales of the 'Third World'', *Third World Quarterly,* 29 (4): 703–21.

Wright, J. (1947) 'Presidential address: terrae incognitae: the place of the imagination in geography', *Annals of the Association of American Geographers,* 37: 1–15.

Yeoh, B. (1996) *Contesting Space: Power Relations and the Urban Built Environment in Colonial Singapore.* Oxford: Oxford University Press.

Young, I.M. (2003) 'The logic of masculinist protection: reflections on the current security state', *Signs: Journal of Women in Culture and Society,* 29 (1): 1–25.

Young, R. (1990) *White Mythologies: Writing, History and the West.* London: Routledge.

Zuhur, S. (1992) *Revealing Reveiling: Islamist Gender Ideology in Contemporary Egypt.* State University of New York Press.

INDEX

working classes 37, 40, 42–7
  and culture 52–3
World as Exhibition (Mitchell) 47
world music 135–6
World's Fairs 47–51, 53–4
  the village in 50–1
Wretched of the Earth (Fanon) 123–8

Xala (Ousmane) 136–9

Yeoh, B. 62–3
Young, I.M. 116
Young, R. 146

zones
  frigid 14
  temperate 14
  torrid 14
Zuhur, S. 119

Research Methods Books from SAGE

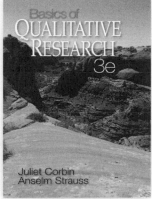

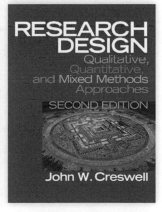

The Qualitative Research Kit

Edited by Uwe Flick

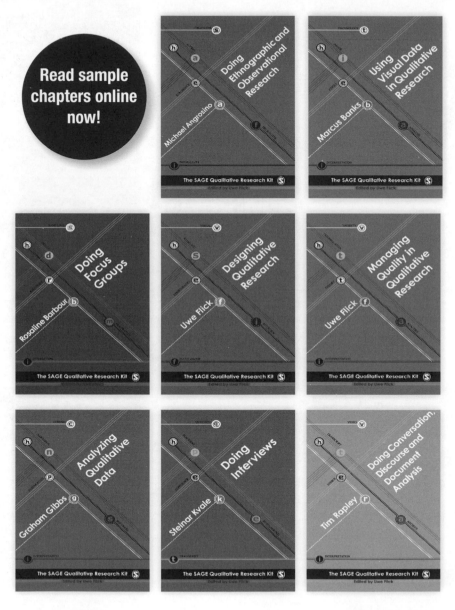

Read sample chapters online now!

Doing Ethnographic and Observational Research — Michael Angrosino — The SAGE Qualitative Research Kit — Edited by Uwe Flick

Using Visual Data in Qualitative Research — Marcus Banks — The SAGE Qualitative Research Kit — Edited by Uwe Flick

Doing Focus Groups — Rosaline Barbour — The SAGE Qualitative Research Kit — Edited by Uwe Flick

Designing Qualitative Research — Uwe Flick — The SAGE Qualitative Research Kit — Edited by Uwe Flick

Managing Quality in Qualitative Research — Uwe Flick — The SAGE Qualitative Research Kit — Edited by Uwe Flick

Analyzing Qualitative Data — Graham Gibbs — The SAGE Qualitative Research Kit — Edited by Uwe Flick

Doing Interviews — Steinar Kvale — The SAGE Qualitative Research Kit — Edited by Uwe Flick

Doing Conversation, Discourse and Document Analysis — Tim Rapley — The SAGE Qualitative Research Kit — Edited by Uwe Flick

www.sagepub.co.uk